Jan Reichl

New Knowledge: The World is composed of whole numbers

2014

New Knowledge: The World is composed
of whole numbers

© Jan Reichl 2014

Address:
Florentinerstrasse 20/2071
D-70619 Stuttgart

janreichl@augustinum.net

The Book is available as E-Book and as
printed book, also in German language:

Neues Wissen: Die Welt ist aufgebaut
aus ganzen Zahlen

Manufactured and published by
BoD - Books on Demand, Norderstedt, Germany

ISBN 978-3-7357-6078-4

The Author was born on 8. September 1931 in Martin, Slovakia. He studied at the Agricultural University in Brno. At the University of California in Davis and at the University of Hohenheim in Stuttgart he developed models for metabolic processes. In years 1979 - 1995 he was professor at the University of Hohenheim. The method of ten steps which he developed connect wide relations between spiritual and natural processes in one unity. The spiritual content of the ten numbers define the processes. Numbers in geometrical figure define performance triangles.

Spiritual content of numbers include also ten synonyms for God. From the synonyms is be possible to deduce all. The synonyms are also basis for health keeping with thoughts. The man as spiritual being is permanently healthy. This can everybody program for himself. Also permanent spiritual love can be programmed, when in thoughts every body is seen to be good and perfect. Then is possible to relax and see everything from above.

E-03

Pages   **Contents**

The world is composed of whole numbers

Pythagoras (572-497 B.C) - Discovery of rational proportions of numbers in the nature lead him to the teaching that the being of the reality is the number. In music is known his law: the shorter the length of a string, the higher its pitch, and vice versa.

Kepler (1571-1630) - The universe is filled with harmony, but harmonic sounds they only when two tones form exact angle of a geometrical figure. He compared music intervals with aspects of planets:

1/2 = 0.5 octave	C-C	conjunction	360 = 360/1
2/3 = 0.67 fifth	C-G	opposition	180 = 360/2
3/4 = 0.75 fourth	C-F	trine	120 = 360/3
4/5 = 0.8 third	C-E	square	90 = 360/4

Intervals: basic tone / specific tone (Y)

	Hz	string C C/Y	string D D/Y	string G G/Y
C´	262	1		
Db	277	16/17		
D	294	8/9	1	
Eb	311	5/6	16/17	
E	330	4/5	5/6	
Gb	370	2.4/3.4	4/5	
G	392	2/3	3/4	1
Ab	415	1.7/2.7		16/17
A	440	1.5/2.5		8/9
Bb	466	1.3/2,3		5/6
B=H	494	1.1/2.1		4/5
C´´	523	1/2		4/4

Not all intervals are whole numbers. This is the reason, why an music instrument has more strings.

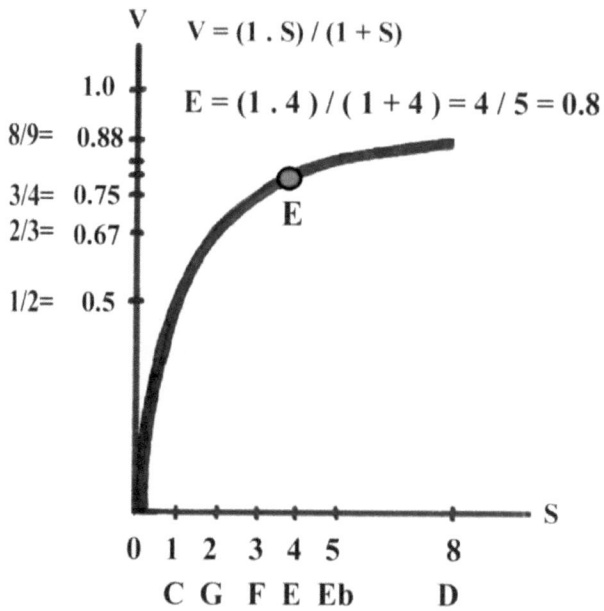

$$V = (1 . S) / (1 + S)$$

$$E = (1 . 4) / (1 + 4) = 4 / 5 = 0.8$$

Not all relations between the tones are whole numbers. In a parable whole numbers are produced:

In the figure is the formula for tone E shown.

0.1 / 0.2 = 0.5	= 1/2
0.2 / 0.3 = 0.667	= 2/3
0.3 / 0.4 = 0.75	= 3/4
0.4 / 0.5 = 0.8	= 4/5
0.5 / 0.6 = 0.833	= 5/6
0.6 / 0.7 = 0.857	= 6/7

Division of 360 degree circle by the number 5 (36, 72 ...) and the number 6 (30, 60) is complementary:

36 / 360 = 0.1 = 30 / 300 = 1 / 10
72 / 360 = 0.2 = 60 / 300 = 1 / 5

E-06

Spiritual content of the first ten numbers

Agrippa von Nettesheim (1486-1535) defined the content of numbers as follows: 1 - origin, 2 - connection and division, 3 - triangle of the wisdom, 4 - duration and strength, 5 - justice, 6 - construction and work, 7 - life, 8 - balance, 9 - activity, 10 - wholeness.
Numbers higher than ten can be reduced by cross summation:

Example: cross sum of 20 and of 11 is 2
             cross sum of 17 and of 26 is 8

.

E-07

New definition of the first ten numbers
(Reichl 2006, 2011)

1	identity	principle	consciousness	mind
2	realization	function	intuition	spirit
3	conditions	factors	thinking	soul
4	demarcation	object	matter	person
5	decomposition	fight	regeneration	life
6	construction	performance	regulation	truth
7	turn	release	environment	love
8	equilibrium	harmony	balance	protection
9	creation	movement	technique	wisdom
0	connection	unity	organization	wholeness

In this sequence are running all processes off. Three numbers can build performance triangles. The terms in the first column are used for description of the processes in this book. The terms in the last column are the synonyms for God. From the synonyms can be everything derived.

Lets start with Ten - step model of Love. Then will be shown four examples of the Method of ten steps. The definition of ten Synonyms for God is used as basis for Health Keeping with Thoughts.

E-08

Ten - step model of love

1. Identity
Universal love exists independently of human beings
2. Realization
This love can be received and given to various relations
3. Conditions
Love can be applied to people, to things, to values
4. Demarcation
Love to one person, or marriage, is demarcation, treaty
5. Decomposition
Demarcated love require faith, wake up jealousy
6. Construction
The rules for common life are formulated
7. Turn
Through release we receive again the universal love
8. Balance
The received love will be given also to other people
9. Creation
Methods for helping to all people will be developed
10. Wholeness
Embrace everybody to the universal love

System of skiing

1	Identity, principle	**ski, skier, snow**
2	Realization, function	**downhill movement on the snow**
3	Conditions, factors	**reacting to the hill and snow**
4	Demarcation, object	**selection of the style**
5	Decomposition, development	**improving the movements**
6	Performance, regulation	**competition skiing**
7	Turn, environment	**free movement in the land**
8	Equilibrium, balance	**balancing the results**
9	Creation, technique	**developing new strategy**
0	Wholeness, unity	**preparing new cycle**

System of business decisions

1	Identity	**recognition of the problem**
2	Realization	**collecting information**
3	Conditions	**alternative solutions**
4	Demarcation	**selection of one alternative**
5	Decomposition	**developing process**
6	Construction	**comparison of results with demands**
7	Turn	**decision for final solution**
8	Equilibrium	**evaluation of sales**
9	Creation	**creative separation from usual competitors**
0	Wholeness	**storing the knowledge, preparing new project**

Formation of a community

0 Start   **thoughts about formation of a community**
1 Identity   **definition of a common language**
2 Realization   **increase of the number of members**
3 Conditions   **occupation of a territory**
4 Demarcation   **marking of borders**
5 Decomposition **fight for enlargement of the territory**
6 Construction   **development of housing**
7 Turn   **changes of the community structure**
8 Equilibrium   **balance of power in the community**
9 Creation   **development of new strategies**
0 Wholeness   **beginning of a new cycle**

The surface of the **computer software**
will be larger **in dependency of ...**

1 Number of names, which can be see on the surface **(Identity)**
2 Modes of relationships between these names   **(Realization)**
3 Sequence of individual operations   **(Conditions)**
4 Communication between the domains   **(Demarcation)**
5 Responsibility to scratch the objects   **(Decomposition)**
6 Competition of tasks during the time   **(Performance)**
7 Simplicity of working with the software   **(Environment)**
8 Evaluation of sales of the software   **(Balance)**
9 Improvement of the software   **(Creation)**
0 Storing the knowledge   **(Wholeness)**

Ten Synonyms for God

In the last column on the page 8 are the synonyms for God presented. From it follows that synonyms could be used for deriving everything in the world and the synonyms could also be used for keeping health with thoughts.

1. **Mind - Identity, Principle**
    Exod 3:14 I am the I AM
    John 1:1 At the beginning was the Word, and the Word was God
2. **Spirit - Realization, Intuition**
    1.Kor 3:16 Know ye not ... That the Spirit of God dwell in you ?
3. **Soul - Conditions, Thinking**
    Acts 4:32 The multitude ... Were of one heart and of one soul
4. **Person - Demarcation, Body**
    Matt 6:9 Our Father which art in heaven
5. **Life - Decomposition, Regeneration**
    John 14:6 I am the way, the truth, and the life
6. **Truth - Construction, Regulation**
    1 John 5:6 The Spirit is truth
7. **Love - Turn, Release**
    1 John 4:16 God is love
8. **Protection - Equilibrium, Existence**
    Isaiah 14:14 I will be like the most High
9. **Wisdom - Creation, Strategy**
    James 3:17 But the wisdom that is from above
10. **Wholeness - Unity, Connection**
    1 Corinth 12:6 It is the same God which work all in all

Man realizes himself in following steps

Steps 1-2-3
In first three steps man collects informations about himself, he is looking for his field of activity und is collecting knowledge about how to manage the tasks.

1. Identity, Principle. The substance of everything is spiritual. In the first step all people are perfect and can develop in any direction.

2. Realization, Intuition. Second step is decision to go in some specific direction. Using intuition man can feel if he is going the right way.

3. Conditions, Thinking. The soul is the interface between spirit and body. In crisis situations are the pain, pleasure and sorrow become conscious.

Steps 4-5-6
In these steps man is developing his personality und he is identifing himself with his body. He try, if necessary, also with fight, put himself through.

4. Demarcation, Person. In the fourth step man demarcates himself from the other people and develops his personal style of appearance.

5. Decomposition, Development. Through decomposition of old and formation of new demarcations man develops.

6. Performance, Body. Truth is the manager that keeps the man on the narrow road to the goal. That increases efficiency of his work.

Steps 7-8-9
In these steps man if he should see everything from above, rise up and then is able to arrange everything clearly.

7. Release, Turn. Through release and rise above the problems, man opens the door to universal love, which he receives and gives further.

8. Balance, Equilibrium. Receiving and giving the universal energy and energy taken from the food, can be programmed.

9.Creation, Strategy. Man creates methods, solutions and strategies for activities and his movement.

Step 10 is Wholeness, Beginning. With integration of knowledge into the wholeness, one Cycle is closed and a new begins.

After closing the cycle man should know what him tied up in his development. Almost all problems start at the step four, when the man demarcate. It could cause problems with his body, but also binding to the self built duties and directions can make problems. Man should learn to rise above the problems, while only then he can receive the uneversal love. Sentence "Love your fellow like yourself" is bad formulated. Man receive universal love and give it with smile and helping further. Also when God will be recognized only as a person, not as universal being, it could make problems. Ten steps of Health Keeping with Thoughts are in detail described.

1. Identity - I am the I AM - Mind

At the spiritual level are all people permanent healthy. Sickness is absence of perfection. Good eliminates evil, when it fills the whole room. Darkness disappears when we lighten the room.

By man developed device works and it is possible to repair it only, because the project where it is described, spiritually exists.

When Jesus said, stand up, you are healthy, he realized the principle, than man is permanently healthy.

At the first step we turn to God. We do not pray to God as to a person, we should adore the principle which is manifested in ten synonyms for God. With the first sentence "Our Father" we address the God. With the second sentence "Thy kingdom come" we opens to his action. With the third sentence "Thy will be done" we rely on his action. It is not necessary to speak out our wishes. Instead of it, we should deepens and listen what we should do.

If some parts of our body do not act correctly, it does not means that we are sick. It means only that some part of the body does not work. We should not deny the principle, that we are permanently healthy.

The instrument of the first step: Knowledge.

2. Realization - Function, Intuition - Spirit

When a man decide to use a certain way, he can prove it by intuition. Perception with senses or extrasensory perception is like a movement in two connected vessels. When is suppressed perception by senses, the man can perceive what the senses are not able to perceive.

Simple method for proving, if a decision is correct, is to define in mind all alternatives, before he falls asleep. During awakening we can prove the alternatives by projecting them in our mind. The alternative which is exciting is not correct, because it is what want the senses. The alternative which is neutral is also not correct. Only when the alternative produce in the man quiet soft harmony, this is the best.

The past, the presence and the future do exists jointly. With senses can be perceived only what is at the moment in our surroundings. Jose Silva (1914-1999) developed a method for transforming consciousness from sensory (beta brain waves) to extrasensory (alpha brain waves) perception by counting. Contact to the spiritual world, which is spatial and temporal indefinite, realized also the prophets.

Prayer is the turning of man to God. Prayer can be spoken loudly, or man can sink into prayer. True prayer is not begging, but the reality affirming power call, formed by knowledge, that God knows what I need before I start to ask.

The instrument of the second step: Intuition.

3. Conditions - Pleasure, Sorrow, Thinking - Soul

Soul is the interface between the spirit and body. Development of self is realized in borderline situations of the life: pain, sorrow, death, pleasure, fear, love.

Filtering of negative emotions in borderline situations is similar to the washing of dirty particles in a layer of wash-active substances. After removing negative emotions, it is necessary put self-confidence to the soul, like a softening agent to the wash article.

Psychosynthesis developed methods which lead thinking to the sources: questionnaire, music playing and listening, free drawing, dancing, free movement, meditation, acting as if, and so on.

One of these methods is the technique of ideal pictures (Assagiol, 1984): 1. What we believe we are, 2. What we would be gladly be, 3. What we would be glad that other people believe we are, 4. What other people believe that we are, 5. What other people want us to be, 6. The picture about us that other people generate in us, 7. Pictures of what we would like to be able to be.

Pain protect the body against injury. Each body has own pain memory. Pain is very closely bound to the social relations. Pain lead to crying, crying calls the mother, the mother takes away the pain. Pain has a connection also to aggression and power. Pain ceases when man rise and thinks about something else.

Instrument of the third step: Thinking.

4. Demarcation - Matter, Body - Person

At the fourth step man identify himself with his body. The association of single molecules to larger aggregates increases the mass. The aggregate gains its own identity and demarcates itself from other aggregates, it forms a surface. Because of gravity, aggregates form a sphere or a globule. Man defining his personality, demarcate from other people and demarcated person become a mass. Communication between the people is like the metabolism in body. The views are losing or winning arguments and stabilize at certain point. We can talk about equilibria like in the metabolism.

After the method of steps the healing is making the mass of disease smaller. In our eyes we should not see a sick man, but a healthy man. From him we separate in our thoughts sickness and bad qualities of his behavior. The infection disease can be also influenced by thoughts. It is like a water falling from sky when the mass of water become large. The sun separate the molecules of water, the water evaporate and disappears.

The healing at the fourth step must start at the first step:
1. **Identity.** There are not good and bad people
2. **Realization.** There are only people which behave badly
3. **Conditions.** Do not fight people which behave badly
4. **Demarcation.** Separate in thought bad qualities from them
5. **Decomposition.** Do not react while they are aggressive
6. **Feedback.** Show them understanding when they are quiet
7. **Turn.** Promise that you will consider they arguments

The Instrument of the fourth step: Communication.

5. Decomposition - Development, Will - Life

Life is generally composition and decomposition, birth and death. In living organisms, cells die continuously. After decomposition, new parts of the body compose again. The psychic life of man develop in six-years periods. Crisis happens when the man did not gain control over the changes.

Development of organisms in steps:

1. **Identity.** The variability is the principle of the development
2. **Realization.** Division of cells brings new forms of cells
3. **Conditions.** Organisms differentiate through adaptation
4. **Demarcation.** Populations need large space for living
5. **Decomposition.** Death and birth produce large variations
6. **Regulation.** Steady state uses minimal energy
7. **Turn.** In a new cycle cooperate different species
8. **Balance.** Equilibria between single communities develop
9. **Creation.** Strategies develop both prey and predator
10. **Wholeness.** Man protects and influences the nature

By killing the bodily being, it is not possible to kill the spiritual being, but killing removes him the possibility for his development. This is a moral question. The moral develops every individual individually with help of his conscience.

Allow not that decomposition of your body will be greater than regeneration. Regenerate your body by various activities. The will is resoluteness, desire. Everything should be done through free will without compulsion.

The instrument of the fifth step: The will.

6. Performance - Construction, Regulation - Truth

Truth is agreement between knowledge and the thing.
It is feedback which hold us on the narrow way to the
goal. In the refrigerator is also feedback which keeps
temperature within narrow bounds.

The sixth step are also religious dogmas, prescriptions
and commands. Commands of Moses are rules for life
in the community, which were written after the exodus
from Egypt.

Dogma is theorem, conviction, which is not through
evidence but through authority defined. In churches it
is revelation, written in Veda, Thora, Bible or Koran.
The churches in the past have realized the obeying of
the scripture also with punishment. At the presence it
practice religious extremists.

In the Age of Enlightenment were natural sciences and
rational reason basis for criticism of irrational ways
of thinking. However, materialism, idealism or dualism
are only abstractions, which do not describe the reality.
Method of ten steps shows, how the spiritual substance
during the steps materialize and further develop.
Only through release and rise above the dogmas and
prescriptions man can receive the universal love, which
solve all problems.

The instrument of the sixth step: The Truth.

7. Turn - Release, Rise - Love

The term love has various meanings. Sexual demands, sympathy, partnership, charity. Passionate love can be opposition of love, therefore the definition of love should not be dependent on relationships. Universal love is love of God, which fills the universe. Universal love exists independently of human beings. We receive this love and give it further. Love to one person is demarcated love, which implies faith and can lead to jealousy.

Love is not desire, love is release. Healing with love needs give up fear about illness. It can be learned. Analogy: When we learn to swim, it requires courage to put ourselves straight into the water. Even without movement of arms and legs, it is possible to stay above the water. The movements of arms and legs can be fully used for movement forward. On the other hand, if we because of fear stretch the head upwards, the legs move down and the movement of arms and legs is fully used for keeping the body above the water.

1. **Identity.** Release, because God is everything
2. **Realization.** When God is love, he love his creation
3. **Conditions.** Do not be worry that for you is no solution
4. **Person.** You have love also when you cannot share it
5. **Decomposition.** Problem disappear not, if you combat it
6. **Performance.** Trust keeps the doors for love open
7. **Turn.** Receiving love exclude everything negative

The instrument of the seventh step: The Release.

8. Equilibrium - Balance, Harmony - Protection

When doctor prescribe a medicament, he is programming patient for action of the medicament. Permanent health can also be programmed, when the man is seen in his spiritual substance as permanently perfect and healthy. This program protect the man when some endanger him. Even when doctor creates the image of the illness, the patient should know, that symptoms can be seen on the body, but the healing starts on a spiritual level first.

Also programming love is for permanent rising about problems. If in the subconsciousness is a program, that everybody as a spiritual being is good, this make the man relaxed. He can see everything from higher point of view in any situation.

Programming of cooperation in politics is about forgetting the right and left political programs, replacing them by universal political program, which automatically finds a way for cooperation. It is about not fighting for power. Instead of fight, praise the opponent when he proposes what is generally good.

Programmed can be also short-term goals, for example if the man want to buy something what he at the moment can not receive in the shop. The man receive to his mind a signal, when he should go to visit the shop. It is to be at the right time at the right place.

The instrument for the eight step: Programming.

9. Creation - Technique, Movement - Wisdom

Healing with energies is predominantly healing with the movement of the body. Movement distributes the energy proportionally in the body. From metabolism in the body man receive primary energies. With movement of body there is induced an positive electromagnetic field, which can be used for healing or for fight.

Proving if the man has healing hands with strong positive energy can be tested on the flat of left hand by man or flat of the right hand by woman. The radiation of the flat turns to left. If the radiation of fingers turn to right, the man has strong positive energy, which can be used for giving the healing energies to places on the body, which have lack of energy.

With Tai Chi movement it is possible to demonstrate the flow of energies between earth and air, which flows through the man. By deep rooting and straight direction of fingers of hands it is not possible to move the man from his position on the ground. Dancing move also energies of the man. The latino swaying hip moves the energies around the body in form of eight. The vertical sway pulls the cosmic energies to the body, stamping steps pull the energies from the earth to the body.

The instrument of the ninth step: The movement.

10. Unification - Connection, Fullness - Wholeness

Wholeness is something which is not defined through its components, but though the connection of the components, for example an organism or a work of art. It is spiritual substance, information. Spiritual healing is grasping of spiritual substance by thoughts.

Meditation is a technique of the control of thoughts, a feeling of quiet peace. Buddhist meditation directs the whole attention to one point or to one's own breath. The goal is the directing of thoughts more and more to the inside, until thoughts self transcend, that means, the border of experience goes above sensory perception. Autogenous training or alpha-wave training are not going so deep, but can also bring positive results. Spiritually developed man can meditate in every situation, also in the mid of mass of people. The ten step models are good frame for seeing things in their wholeness.

Masses of people build also a wholeness. The antique Greeks turned fights between the cities to competition in Olympic games. Some groupings of people have violating acts in their programs. On the other side, in mass concerts the disproportion between small number of acting people and large mass of observers is good managed by music.

The instrument of the tenth step: The Gratitude.

The "Lords Prayer" (Matthew 6:9) corresponds to first seven steps of keeping healthy with thoughts:

1. Identity: Our Father which art in heaven
The address, not My Father, but Our Father, express love

2. Realization: Thy kingdom come
It is opening to Gods action, an affirmation, not request

3. Conditions: Thy will be done in earth as it is in heaven
God's will is health and fullness, we rely on his will

4. Demarcation: Give us this day our daily bread
Not my bread, but our bread

5. Decomposition: And forgive us our debts, as we forgive ...
Our will can cause wrong, therefore forgive ...

6. Performance: And lead us not in temptation ...
Direct way controlled by truth is straight and narrow

7. Release: For thine is the kingdom and the power ...
When I am in the Kingdom of God, I can release

First seven chapters of Koran follow also method of steps
1. Identity 1:14 You alone we ask for help
2. Realization 2:2 The book is directional cord for god-fearing
3. Conditions 3:14 Embellished are the people the love ...
4. Persons 4:36 Goodness to parents, relatives, orphans ...
5. Life 5:3 Forbidden is you the ... choked, killed ...
6. Truth 6:32 The dwelling in glory for the honest
7. Love 7:24 He spoke to Adam and Eve: go out from here, where one to other enemy is

Age progress: 6 years / step

Small red numbers are steps - uneven numbers are on peaks, even numbers in dales of the pentagram. Large black numbers are the age. Below is the age progress presented by the curve of half-circles. Culmination in the low part of curve corresponds to the even step numbers in low points of the pentagram. Culmination in the upper part of the curve corresponds to the uneven step numbers on peaks of the pentagram. Break points between the half-circles are turning points between the periods.

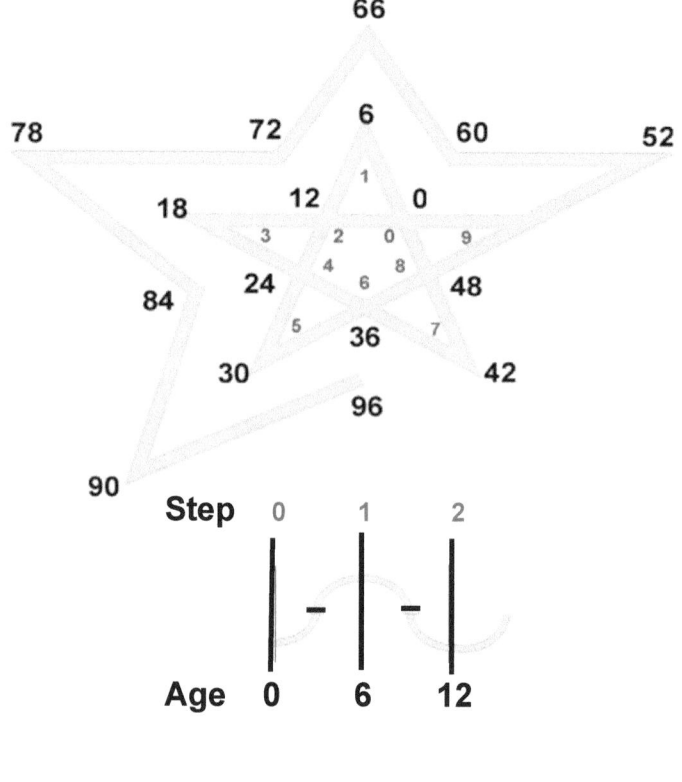

Development of man in 6- year periods

Age   Step
0-6   0-**1 (Identity)** collection of knowledge about self
6-12   1-**2 (Realization)** finding the place for own actions
12-18   2-**3 (Conditions)** puberty, professional education
18-24   3-**4 (Demarcation)** studies, leaving the parents house
24-30   4-**5 (Decomposition)** experimentation, own existence
30-36   5-**6 (Performance)** finding permanent job
36-42   6-**7 (Turn)** finding new partner relationships
42-48   7-**8 (Equilibrium)** deeper changes in self
48-54   8-**9 (Creation)** new life philosophy
54-60   9-**0 (Wholeness)** peak of success in profession
60-66   0-**1 (Identity)** looking for new activities
66-72   1-**2 (Realization)** realization of new activities
72-78   2-**3 (Conditions)** submit the experiences to others
78-84   3-**4 (Demarcation)** recognize of own limits
84-90   4-**5 (Decomposition)** regeneration of body activities
90-96   5-**6 (Performance)** rely on the spiritual principle

Turning points - crisis years

Age

3 - childhood crisis   **6 - reality crisis**
15 - puberty   **21 - crisis of ideals**
27 - self-manifestation   **33 - crisis of career**
39 - status seeking crisis   **45 - middle life crisis**
51 - life philosophy   **57 - authority crisis**
63 - spiritual puberty   **69 - new realizations**
75 - hindrances   **81 - comparing with others**
87 - self-healing   **93 - deepening**

The 6-years age progress developed Huber (1930-1999).

Age progress
6 years / one House
after the method of
Huber (1980)

With this method it is
possible examine also
action of planets on
the age progress

Sun - consciousness
Moon - emotions
Mercury - communication
Venus - beauty
Mars - sport performances
Saturn - stability
Jupiter - social status
Uranus - spirituality
Neptune - arts
Pluto - business

Additional actions:
Mars - pregnancy
Saturn - wedding
Uranus - separation

The turn point from
one house to next
is like in the method
of pentagram in
the middle up to
second third of the
house. Maximum
of the action is at the
start of the house. Sun

54
MC   48
60   42
66   DC 36
0/72   30
6/78   24
12/84 IC 18/90

at start of the house shows an extroverted man.
Sun in the low point of house shows an introverted
man. At low point could occur crisis.
The age is written outside of the circle at the start
of the house.

Definition of the signs and houses with method of steps
x) definition of Amer. Astrol. Federat. (Goodavage, 1968)

1 Identity, Subject	Aries (1.House)	I am x)
2 Realization, Possession	Taurus (2.H.)	I have
3 Conditions, Learning	Gemini (3.H.)	I think
4 Demarcation, Home	Cancer (4.H.)	I feel
5 Regeneration, Life	Leo (5.H.)	I will
6 Performance, System	Virgo (6.H.)	I analyze
7 Environment, Partner	Libra (7.H.)	I give
8 Giving, Taking	Scorpio (8.H)	I desire
9. Creation, Critics	Sagittarius (9.H.)	I see
10 Organize, Connect	Capricorn (10.H.)	I use
11 Identity, Knowing	Aquarius (11.H.)	I know
12 Realization, Feeling	Pisces (12.H.)	I believe

Integration of the oppositions in astrological cycle
1 **Aries** concentrate on himself
7 **Libra** need a partner
2 **Taurus** want owe also his partner
8 **Scorpio** gives freedom but is jealous
3 **Gemini** learn and is not sure to talk about
9 **Sagittarius** talks at all times also when is nothing to say
4 **Cancer** feels with collective
10 **Capricorn** manager with no contacts to individuals
5 **Leo** has his own moral *To be born in a specific*
11 **Aquarius** use ethics *sign, it does not mean*
6 **Virgo** has a system *to be dependent on it*
12 **Pisces** has intuition *through the whole life.*
If man integrates the qualities of the opposite sign, he move to the mid of the horoscope and will be independent on it.

Integration of the qualities in the pentagram
3 Learning - 11 Knowing
4 Demarcation - 10 Connection
5 Decomposition - 9 Creation
6 Construction - 8 Equilibrium

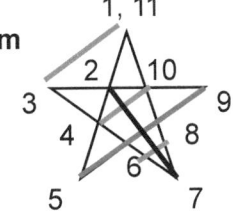

The position of AC (rise of sun) and DC (setting of sun) is determined by the time of the day

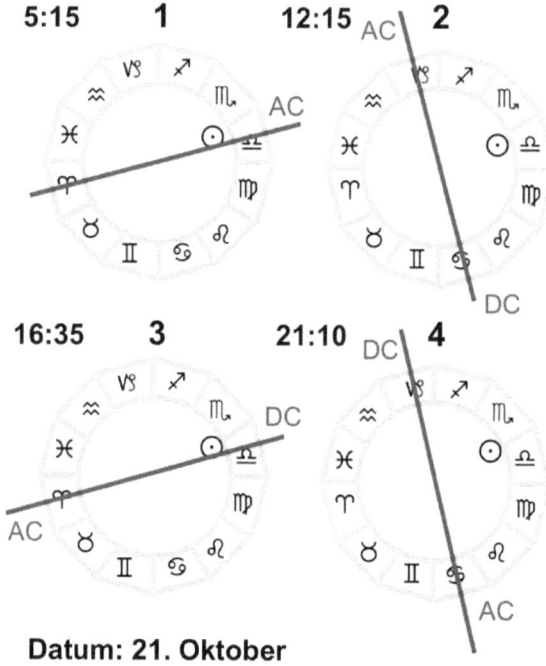

5:15   **1**   12:15 AC **2**

16:35   **3**   21:10 DC **4**

Datum: 21. Oktober

1 - Sun at the side of AC. These Persons are concentrated on themself. They have problem to understand the partner.
2 - Sun above the axis AC-DC. These Persons want to rise above the masses. They have problem to subordinate.
3 - Sun at the side of DC. These Persons live through the partners.
4 - Sun below the axis AC-DC. These Persons seek security in the collective.

E-30

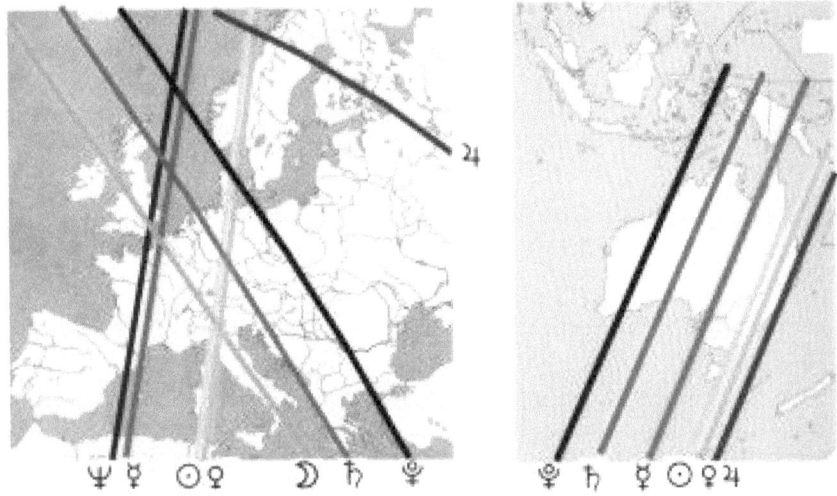

Astro-Carto-Graphy (Lewis 1976) is the projection of axes AC-DC and MC-IC of sun, moon and planets on the world map.

In first example IC-Sun shows the place, where the person settled. The crossing with AC-Saturn shows place, from where the person received support for his work.

Second example shows a person, who as a child said, that she want immigrate to Australia. The person has Sun and many other planes above the Australia.

E-31

Models for body development

are constructed with the method of steps.
They show the unity of matter, energy and spirit.

Statements of materialists and skeptics:
1. It does not exist not material substance which reign the body.

2. All spiritual qualities of man are functions of material body.

3. Declaration of a patient that his condition is improved after some treatment, is from statisticians declared as anecdotal.

4. Also positive result with large number of patients is not enough, it is necessary to have control groups and statistical evaluation.

Answers:
1. One by man developed instrument works and it is possible to repair it, because a project where it is described spiritually exist.

2. The method of steps show how the spiritual identity materially demarcate und further develop.

3. Healing is always individual, first appear it at spiritual level, than materially.

4. Only deviations from the principle not from the mean does have sense and shows the cause of the problem.

The method of steps uses knowledge based models. Statistical methods based on coincidence and deviation from the mean, cannot be used for large models for metabolic processes, neither for long-term forecasts of weather.

Mathematics of ten step models

1 Identity: selection of variables (x, y) and constants (k)
2 Realization: giving the elements in a function (y = x . K)
3 Conditions: system of equation where one determine other
4 Demarcation: limited space determines concentration
5 Decomposition: reversibility and steady states
6 Regulation: feedbacks regulate the system
7 Environment: block diagram of complex relationships
8 Equilibrium: input/output balance determines regulation in 6
9 Creation: strategies determine decomposition in step 5
10 Wholeness: whole evaluation determine the space in step 4

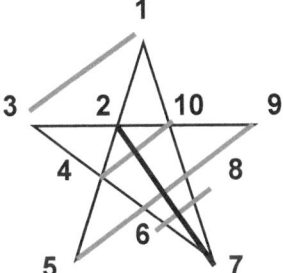

Performance triangles to axis
2 (function) - 7 (environment):

6 regulation   - 8 equilibrium
5 decomposition - 9 creation
4 demarcation   - 10 wholeness
3 conditions   - 1 identity

First will be calculated energy production, blood flow and per cent of protein in whole body of a standard animal, using equation for calculation parameters for a sphere. Then will be shown how steady state adjust and long-term growth of protein content in the body can be calculated. Then block diagram for metabolism will be shown and technique for calculation of rate constants presented.

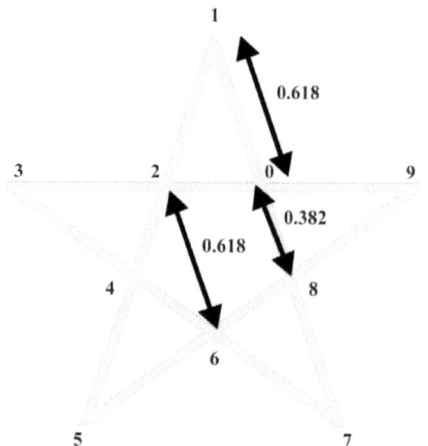

Golden section: 0.618 + 0.382 = 1
0.616 x 0.618 = 0.382

On the peaks of pentagram are uneven male active numbers, in the dales of pentagram are even female restorative numbers. With a series of whole uneven numbers it is possible to define the number "pi/4". With a series of even numbers is defined the basis of natural logarithm "e". With constant "pi" can be calculated heat production, with constant "e" the amount of substances in the body.

Uneven numbers: pi/4 = 0.785 = 1 - 1/3 +1/5 - 1/7 + 1/9 - 1/11 ....
Even numbers   e   = 2.718 = 1 + 1/1 + 1/2 + 1/6 + 1/24 + 1/120 ....

Theoretically defined standard animal with parameters for calculation of sphere: V = volume, A = surface, r = radius, d = diameter

$V = 4.189 \ r \uparrow 3$ 4.189 = 4 pi / 3   =   joule / calorie = kj / kcal
$A = 12.567 \ r \uparrow 2$ 12.567 = 4 pi   =   3 kcal . 4.189 = kj / h
100 / 6 = 16.667 = reaction surface = per cent of protein in the body
A / V = 6 / d   ln 2 = 0.693   18.129 = 4 pi / 0.693

Coulson (1986)

Animal	Weight kg	Blood flow lit/h/kg	Halflife h
rat	0.1	24	0.69
dog	10	6.3	2.64
man	70	4.3	3.86
cow	400	2.8	5.92
lizard	0.005	2.4	6.90
turtle	1	0.7	24.50

Standard Animal

Animal	Weight kg	Flow1 lit/h/kg	Halflife h	A / V $\uparrow$0.75
rat	0.1	22.35	0.75	21.01
dog	10	7.07	2.36	6.68
man	70	4.34	3.84	4.11
cow	400	2.83	5.89	2.66
lizard	0.005	2.87	5.81	2.87
turtle	1	0.76	21.93	0.76

Flow1 = basal blood flow lit/h/kg = 12.567 W $\uparrow$0.25
Halflife1 = basal halflife h = 16.667 / Flow1
Amount of blood lit/kg = 1 / 18.129 = 0.055

Poikilotherm animals: calculated Flow1 divide by 16.47 !!!

Through wide spectrum of homoiotherm and poikilotherm (cold/blooded) animals the calculated values of blood flow and halflife are very close to the measured values.

Adjustment of the steady state flow and of pools at the start and end of the time cycle

$$e^{-1.1} = 0.333$$

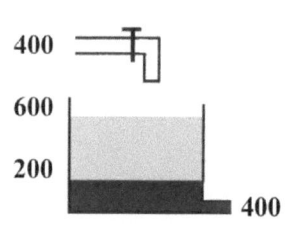

400	......	133
533	......	177
577	......	192
592	......	197
597	......	198
598	......	199
600	......	200

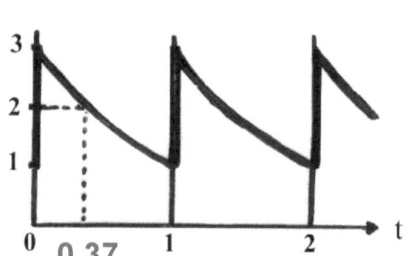

Exp(-1.1) = 0.333
In 0.333 = - 1.1
1 / 0.333 = 3
Halflife = 0.37

Steady state flow is reached, when after repeating of the same inputs at the start of the cycle, the proportions of input, output and flow are whole numbers:

        600 = pool at the start of cycle
        200 = pool at the end of cycle
        400 = steady flow

Calculation of the halflife:
T0.5 = In (exp(-k) + 0.5 (1 - exp(-k))) / -k

E-36

Adjusting of steady state of the growth of microbe

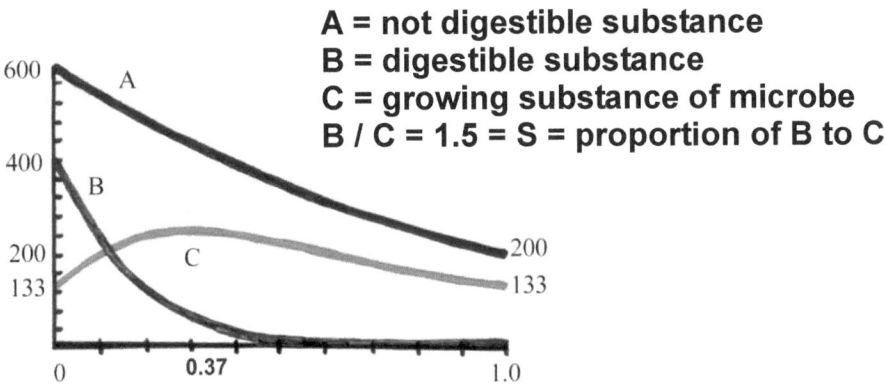

A = not digestible substance
B = digestible substance
C = growing substance of microbe
B / C = 1.5 = S = proportion of B to C

Exp (-1.1) = 0.333

Flow rate constant K1 = =1.1
Growth constant K2 = K1 . 6

Growth constant K2 must be
6 times greater than K1

A = A + (- A . K1) dt
B = B + (- B . K1 - B . C . S . K2) dt
C = C + (- C . K1 + B . C . K2) dt

Endpool of A = A . Exp(-K1) = 600 . 0.333 = 200
Startpool of A = 1 / Exp(-K1) = 200 / 0.333 = 600
Start- and End- Pool of Microbe C = A . Exp(-K1) / S = 133
Startpool of B = Flow of B = 400
Flow of A = 600 - 200 = 400

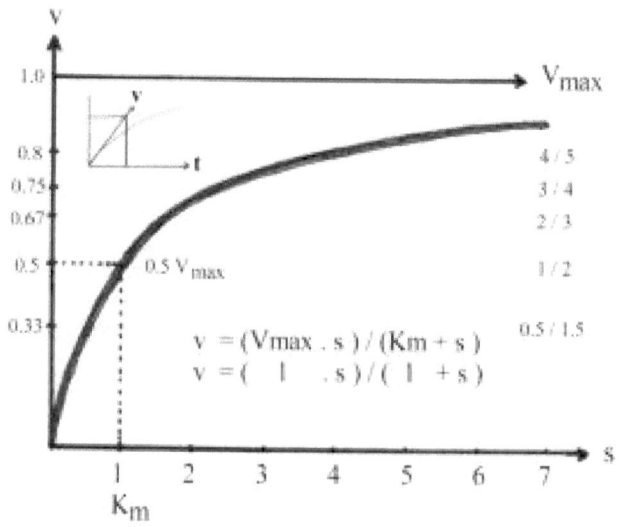

Dependency of the velocity "v" on the amount of the substrate "s". Vmax is the maximum "v". Km is the amount of substrate at the halve Vmax. The curve is a parable 1/2, 2/3, 3/4, 4/5 ...

$$v = (V_{max} \cdot s) / (K_m + s)$$
$$v = (1 \cdot s) / (1 + s)$$

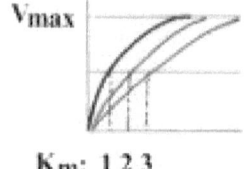

Km: 1 2 3

Competitive inhibition
- different Km

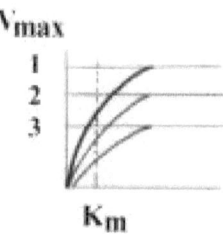

Km

Non-competitive inhibition
- different Vmax

E-38

Daily protein accretion in the animal body

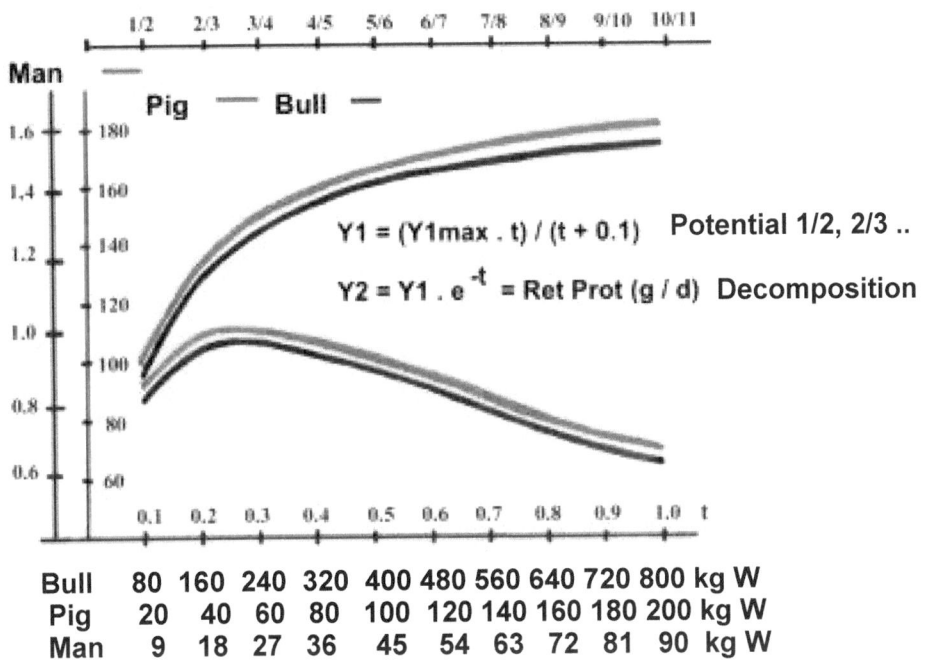

Bull	80	160	240	320	400	480	560	640	720	800	kg W
Pig	20	40	60	80	100	120	140	160	180	200	kg W
Man	9	18	27	36	45	54	63	72	81	90	kg W

Potential protein accretion Y1 in the body (1/2, 2/3, 3/4, 4/5 ..) will change to real daily accretion Y2 (g/d) through decomposition reaction Y2 = Y1 . Exp(-t).

E-39

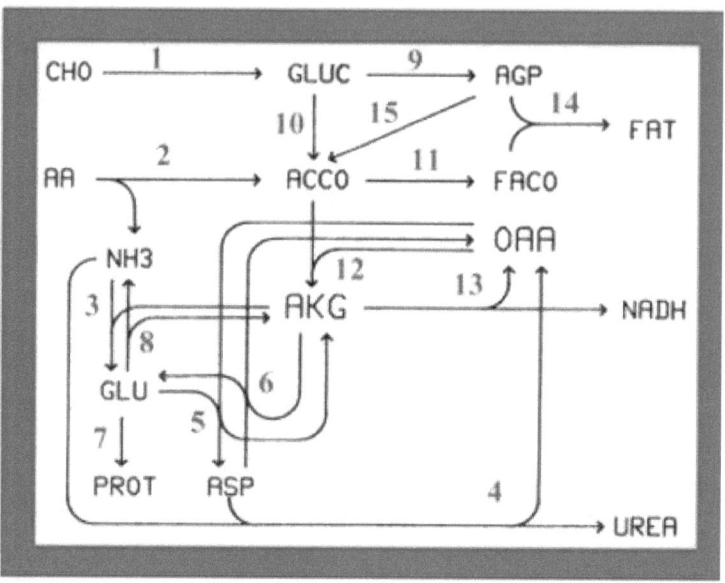

Block diagram of simplified metabolism in the body: The inputs are carbohydrates (CHO) and amino acids (AA). Outputs are protein (PROT) and fat (FAT) in the body. Decomposition products are urea (UREA) and NADH, which is used as energy source for reactions in the body.

The block diagram was constructed using the method of steps:
1. Identity: names of chemicals (CHO, AA, GLUC ...)
2. Realization: metabolic pathways (F01, F02 ...)
3. Conditions: branching and cycling of pathways (R3/R4 ...)
4. Demarcation: synthesis and storage (FAT, PROT)
5. Decomposition: reversibility, transamination (F3/F8, F5/F6)
6. Performance: production of NADH for energetic reactions
7. Release: use of electrical potentials
Individual reactions are in algebraical form presented on the next page.

```
            input-driven sequence equations
               ( R =  branching ratio )
01          - CHO + 1.8 GLUC
02          - AA + ACCO + 1.2 NH3 + 2 NADH
03  R3      - NH3 - AKG + GLU - NADH  ◄─────────┐
04  R4      - NH3 - ASP + OAA + UREA             │
05          - OAA - GLU + ASP + AKG              │
06          - ASP - AKG + GLU + OAA              │
07  R7      - GLU + PROT                         │
08  R8      - GLU + AKG + NH3 + NADH  ── NH3 ────┘
09  R9      - GLUC + 2 AGP - 2 NADH
10  R10     - GLUC + 2 ACCO + 4 NADH  ◄──────────┐
11  R11     - ACCO + 0.14 FACO - 1.8 NADH        │
12  R12     - ACCO - OAA + AKG + NADH            │
13          - AKG + OAA + 3 NADH                 │
14          - FACO - 0.33 AGP + 0.33 FAT         │
15          - AGP + ACCO + 3 NADH  ── ACCO ──────┘
```

Before dynamical simulation begins, the size of fluxes
will be algebraically calculated. R is branching of two
reactions, for example R3/R4 is branching of NH3.
NH3 and ACCO are recycling so long, until their amount
is zero. Then from the size of fluxes the rate constants
can be calculated Flux = K . Pool

TCA (three carbon acid) cycle
(Krebs - Cycle)

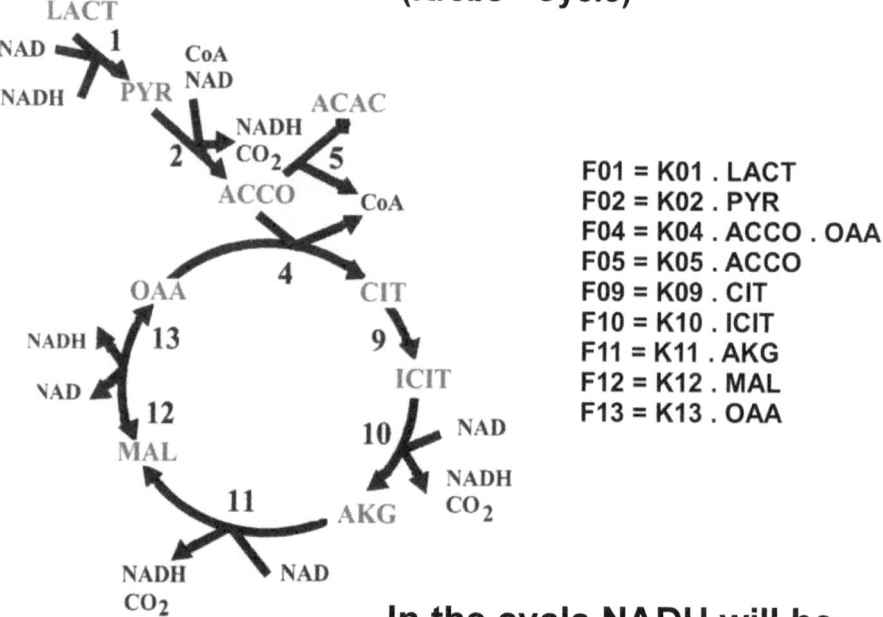

F01 = K01 . LACT
F02 = K02 . PYR
F04 = K04 . ACCO . OAA
F05 = K05 . ACCO
F09 = K09 . CIT
F10 = K10 . ICIT
F11 = K11 . AKG
F12 = K12 . MAL
F13 = K13 . OAA

In the cycle NADH will be produced and then in the respiration chain to H2O oxidized.

With exception of reaction F04, which is reaction of second order, all other reactions are reactions of first order, when only one chemical should be multiplied by the constant.

E-42

In the respiration chain of the mitochondria is from the food primary electrical field produced. Through physical movement of the primary field, secondary electromagnetic field is produced.

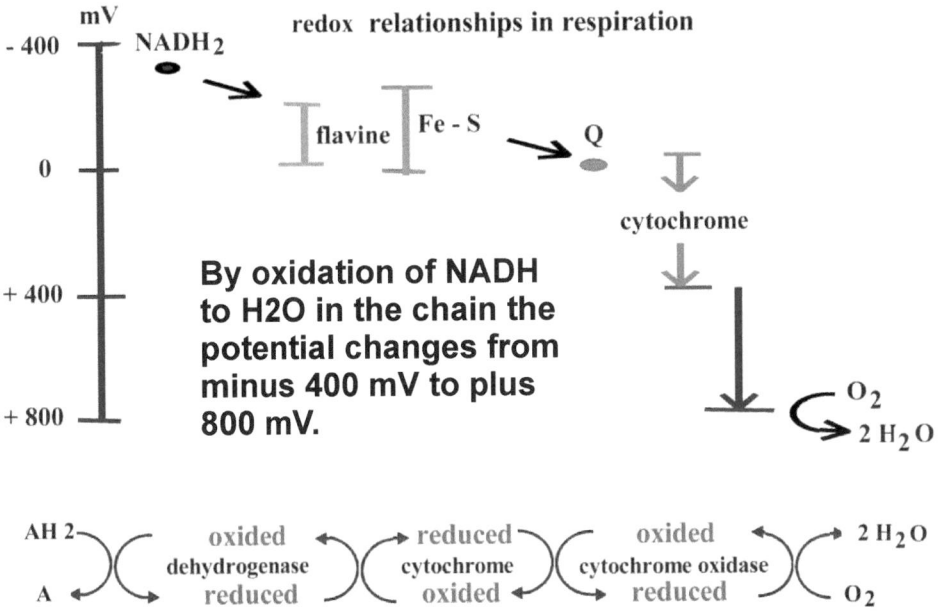

redox relationships in respiration

By oxidation of NADH to H2O in the chain the potential changes from minus 400 mV to plus 800 mV.

E-43

The corkscrew rule of Maxwell shows that left turning current is moving to outside, the right turning move to the inside. The magnetic effect distinguish primary and secondary field. With pendulum it is possible to measure negative left and positive right turning current.

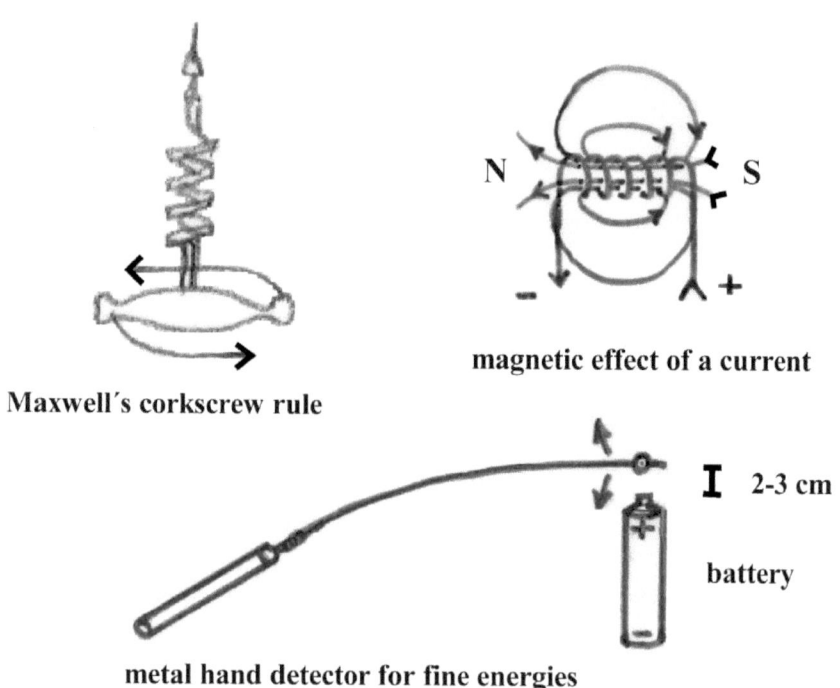

Maxwell´s corkscrew rule

magnetic effect of a current

metal hand detector for fine energies

I 2-3 cm

battery

E-44

Polarity of chakras by
woman   **man**

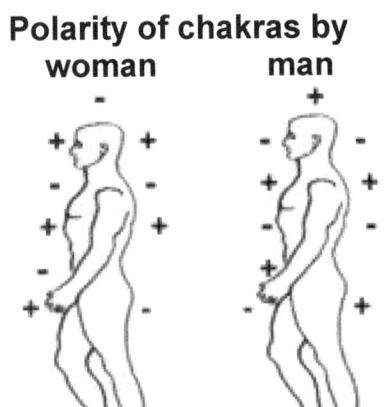

By woman and man is the polarity of chakras opposite

Calculation of energies on the body points from the date of birth

21 10 1980 - 4
56 67 6987 - 9

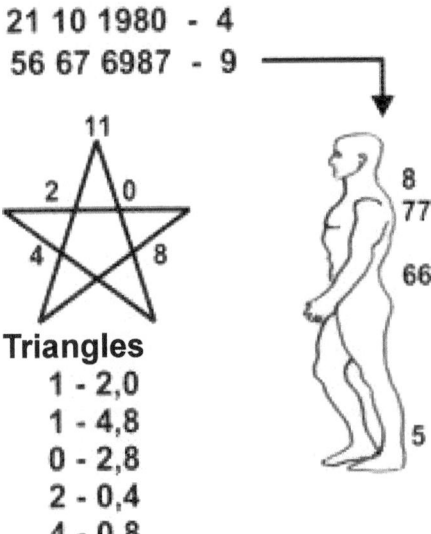

Triangles
1 - 2,0
1 - 4,8
0 - 2,8
2 - 0,4
4 - 0,8

The numbers of the date of birth are to the pentagram directly written, on the body points after correction of numbers 0,1,2 by sum 7 to 7,6,5.
The sum of corrected numbers is showing with which number at the back of head start the writing of numbers on the body points. This method was developed through generations in the Switzerland. I learned it from Fritz Guggisberg (1919-1995).

Numbers corresponding to thousand and hundred are used in the sum, but are not written in pentagram and on body.

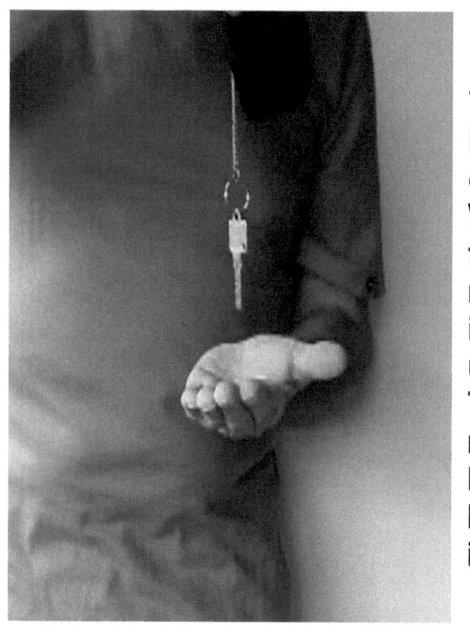

The polarity of heart by man is minus and also at the flat of left hand is polarity minus. When at the end if fingers turn the polarity to plus, the man has strong positive induced energy, which can be used for healing with hands. This test can by woman be made on flat of the right hand. This measurement can be made by a pendulum as it is shown on the picture.

The polarity of the primary energy from the metabolism is minus. The induced energy field which is produced through movement of the primary field, is plus. It is explained by electromagnetic effect of current on page 44. The left turning primary energy is in Tai Chi called Yin, the right turning induced energy is Yang. After Chia and Li (1996) during bending forward and breathing out will be the Yang energy in the stretched forward arms collected and can be used for fight. With proper stand and straight direction of fingers man become deeply rooted and it is not possible move him off the ground.

Projection of the pentagram on the body

 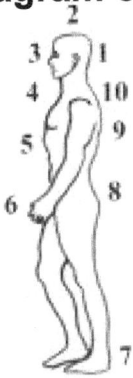

1. Identity:	**Reception**	6. Construction:	**Performance**
2. Realization:	**Inspiration**	7. Turn:	**Moving**
3. Conditions:	**Thinking**	8. Equilibrium:	**Helping**
4. Demarcation:	**Talking**	9. Creation:	**Strategies**
5. Development:	**Willpower**	10. Wholeness:	**Organization**

In this way it is possible to evaluate psychic qualities of a man. The numbers of the birth date activate the body points.

Performance triangles to axis 1(Spirit) - 6(Body)

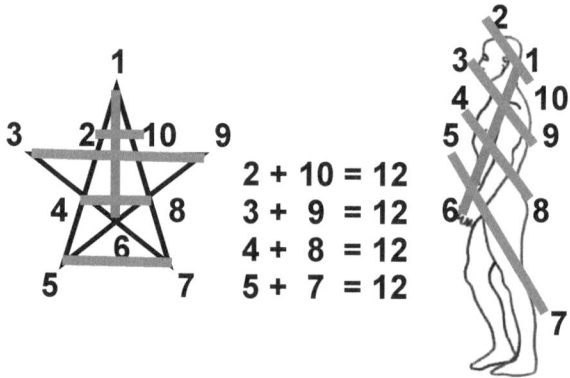

2 + 10 = 12
3 + 9 = 12
4 + 8 = 12
5 + 7 = 12

Triangles related to Point 1 (Spirituality):

1- 2+10 = 12 intuitive, extrasensory perception
1- 3+9 = 12 thinking in wide relations
1- 4+8 = 12 artistic talents
1- 5+7 = 12 strong vitality

Triangles related to Point 6 (Body):

6- 2+10 = 12 magic bodily radiation
6- 3+9 = 12 personal putting through
6- 4+8 = 12 personal sensuality
6- 5+7 = 12 personal showmanship

Performance triangles to axis 2(Korona) - 7(Legs)

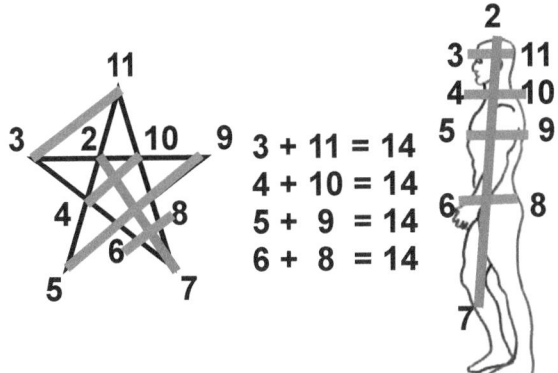

3 + 11 = 14
4 + 10 = 14
5 + 9 = 14
6 + 8 = 14

Axis 2-7 is an energy axis, which connect the energies in the air with the energies in the earth. With the sum 14 will be the energies which are in the front (personal enforcement) refined through the energies which are in the back (helping).

Triangles   front - back
3+11 = 14 Intellectual   (3-thinking, 11-conscionsness) head
4+10 = 14 Organizer   (4-matter,   10-wholeness)   neck
5+9   = 14 Diplomat   (5-will,   9-creation)   breast
6+8   = 14 Striver for power(6- performance, 8-balance) basis

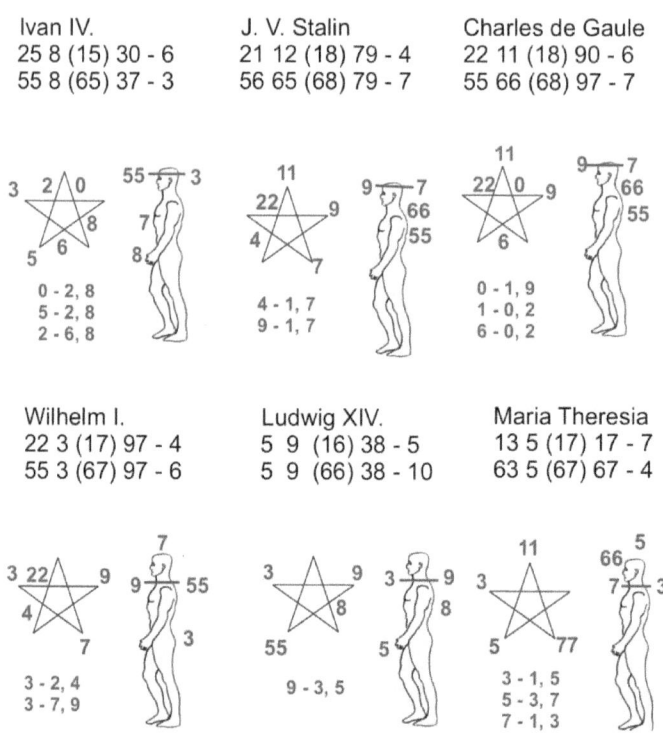

Ivan IV.
25 8 (15) 30 - 6
55 8 (65) 37 - 3

J. V. Stalin
21 12 (18) 79 - 4
56 65 (68) 79 - 7

Charles de Gaule
22 11 (18) 90 - 6
55 66 (68) 97 - 7

0 - 2, 8
5 - 2, 8
2 - 6, 8

4 - 1, 7
9 - 1, 7

0 - 1, 9
1 - 0, 2
6 - 0, 2

Wilhelm I.
22 3 (17) 97 - 4
55 3 (67) 97 - 6

Ludwig XIV.
5 9 (16) 38 - 5
5 9 (66) 38 - 10

Maria Theresia
13 5 (17) 17 - 7
63 5 (67) 67 - 4

3 - 2, 4
3 - 7, 9

9 - 3, 5

3 - 1, 5
5 - 3, 7
7 - 1, 3

Energy axis at the head
Intellectual decisions

Energy axis at the neck
Wholeness decisions

E-50

Konstantin I.
27 2 (2) 80 - 3
57 5 (5) 87 - 10

Adolf Hitler
20 4 (18) 89 - 5
57 4 (68) 89 - 2

J. F. Kennedy
29 5 (19) 17 - 7
59 5 (69) 67 - 2

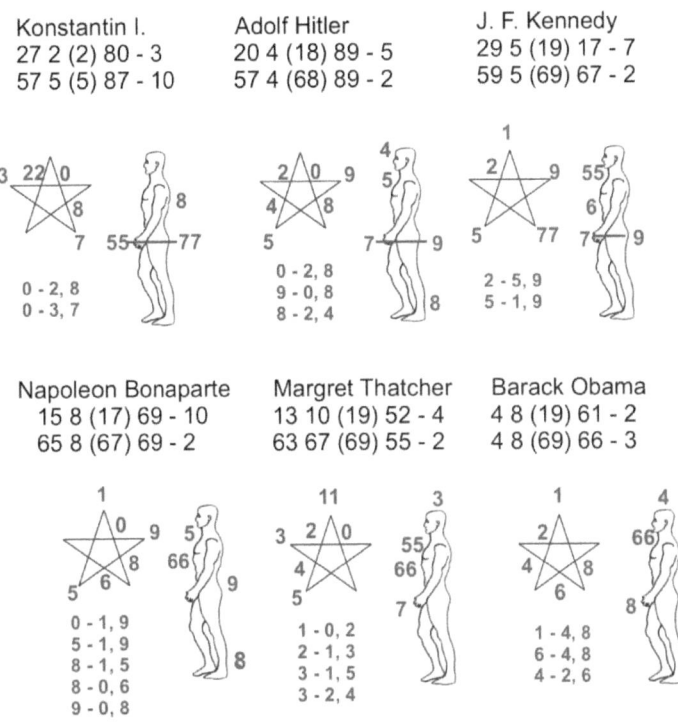

3 22 0
8
7   55   77

0 - 2, 8
0 - 3, 7

2 0   9
4   8
5   7   9

0 - 2, 8
9 - 0, 8
8 - 2, 4

4
5

8

1
2   9   55
6
5   77   7   9

2 - 5, 9
5 - 1, 9

Napoleon Bonaparte
15 8 (17) 69 - 10
65 8 (67) 69 - 2

Margret Thatcher
13 10 (19) 52 - 4
63 67 (69) 55 - 2

Barack Obama
4 8 (19) 61 - 2
4 8 (69) 66 - 3

1
0   9   5
8   66
5   6   9

0 - 1, 9
5 - 1, 9
8 - 1, 5
8 - 0, 6
9 - 0, 8

8

11
3 2 0
4
5   7

3
55
66

1 - 0, 2
2 - 1, 3
3 - 1, 5
3 - 2, 4

1
2
4   8
6

4
66

8

1 - 4, 8
6 - 4, 8
4 - 2, 6

**Energy axis a the bottom of the body
Striving for power**

**Energies at the front of the body
Personal putting through**

E-51

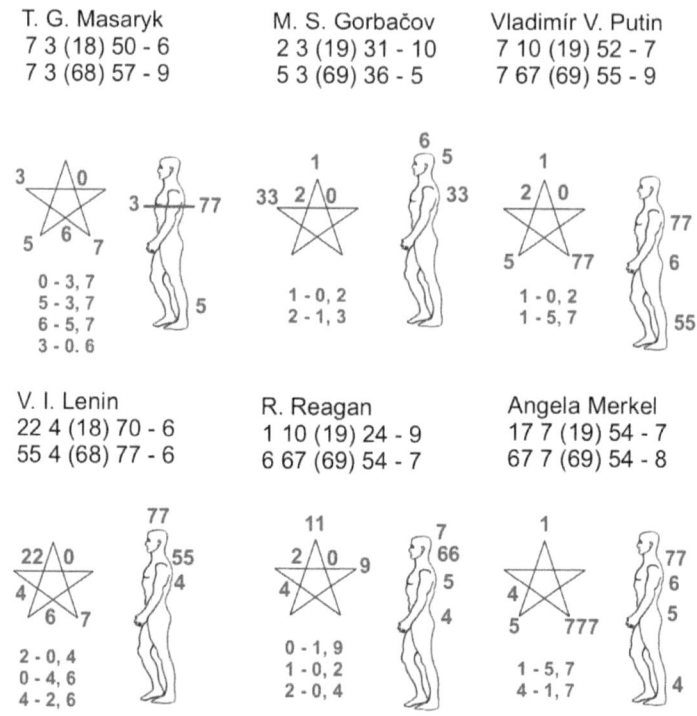

T. G. Masaryk
7 3 (18) 50 - 6
7 3 (68) 57 - 9

3   0
3 — 77
5   6   7
0 - 3, 7
5 - 3, 7
6 - 5, 7
3 - 0. 6
5

M. S. Gorbačov
2 3 (19) 31 - 10
5 3 (69) 36 - 5

6   5
1
33 2   0   33
1 - 0, 2
2 - 1, 3

Vladimír V. Putin
7 10 (19) 52 - 7
7 67 (69) 55 - 9

1
2   0
5   77
1 - 0, 2
1 - 5, 7
77
6
55

V. I. Lenin
22 4 (18) 70 - 6
55 4 (68) 77 - 6

22   0
4
6   7
2 - 0, 4
0 - 4, 6
4 - 2, 6
77
55
4

R. Reagan
1 10 (19) 24 - 9
6 67 (69) 54 - 7

11
2   0   9
4
0 - 1, 9
1 - 0, 2
2 - 0, 4
7
66
5
4

Angela Merkel
17 7 (19) 54 - 7
67 7 (69) 54 - 8

1
4
5   777
1 - 5, 7
4 - 1, 7
77
6
5
4

Energies at the back of the body

Decisions for the whole population

E-52

Periodic System of Elements

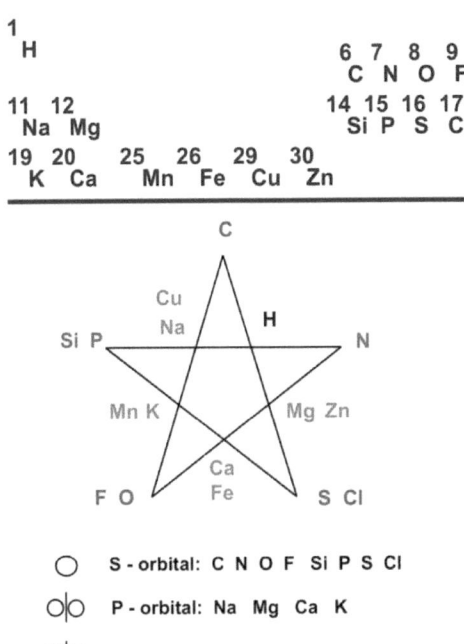

```
1
 H                          6  7  8  9
                            C  N  O  F
11 12                      14 15 16 17
 Na Mg                      Si P  S  Cl
19 20    25 26 29 30
 K  Ca   Mn Fe Cu Zn
```

S - orbital: C N O F Si P S Cl

P - orbital: Na Mg Ca K

D - orbital: Mn Fe Cu Zn

Periods are the horizontal rows. In right direction increase the atomic number of elements. **Groups** are elements in one column. The difference between the rows in group correspond to eight atomic numbers. In the pentagram at the peaks are **donators** of electrons, elements which give energy free. In the dales of pentagram are **acceptors** of electrons, elements which store electrons.

At the peaks of pentagram are donators, in the dales are acceptors of electrons.

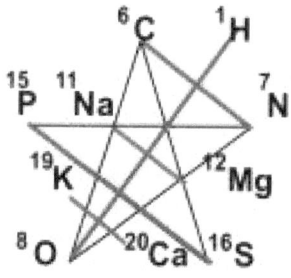

Triangles to axis 5 (O) - 10 (H):

2+8=10: Na-Mg: NaOH, Mg(OH)$_2$
4+6=10: K-Ca: KOH, Ca(OH)$_2$
3+7=10: P-S: H$_3$PO$_4$, H$_2$SO$_4$
1+9=10: C-N: C$_3$H$_7$NO$_2$ alanine

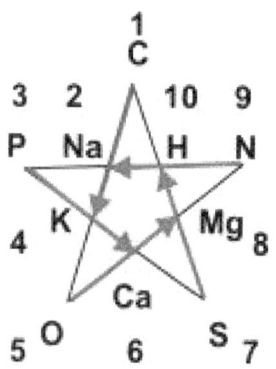

Donator-Acceptor Parallels:

C-K: K$_2$CO$_3$, HCOOK   1+3=4   K
P-Ca: Ca$_3$(PO$_4$)$_2$ 3+3=6   Ca
O-Mg: MgO, Mg(OH)$_2$ 5+3=8   Mg
S-H: H$_2$S, H$_2$SO$_4$ 7+3=10 H
N-Na: NaNO$_3$, NaNH$_3$ 9+3=12 Na

Pentagram Numbers

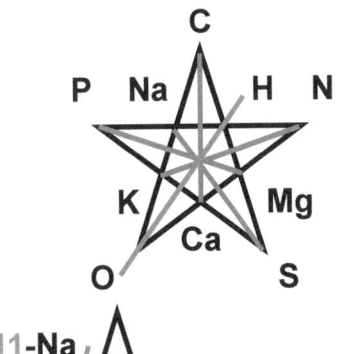

In opposite are Donator - Acceptor

C - Ca	1+5=	6	CaCO3
P - Mg	3+5=	8	Mg-phosphatase
O - H	5+5=	10	H2O
S - Na	7+5=	12	Na2SO4
N - K	9+5=	14	KNO3

Atomic numbers

Exchange: Acceptor - Acceptor
Donator - Donator

K - Na   11+8 = 19-**K**
NaNO3+KCl = KNO3+NaCl
Ca - Mg 12+8 = 20-**Ca**
MgCl2+Ca(OH)2 = Mg(OH)2+CaCl2

P - N   7+8 = 15-**P**
3 P+5 HNO3+2 H2O = 3 H3PO4+5 NO
S - O   8+8 = 16-**S**
H2S+1/2 O2 = S+H2O

E-55

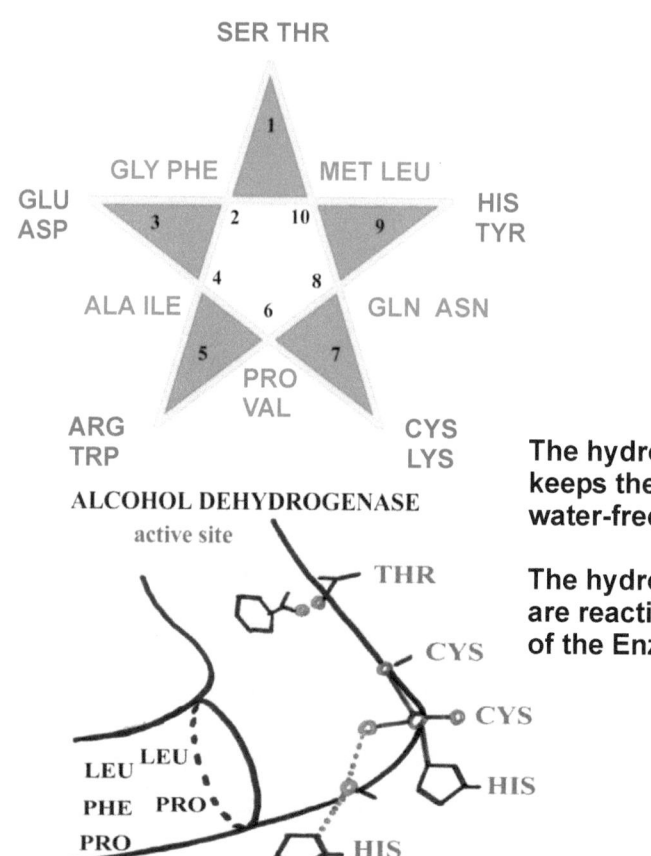

SER THR

GLY PHE MET LEU

GLU
ASP

HIS
TYR

ALA ILE

GLN ASN

PRO
VAL

ARG
TRP

CYS
LYS

ALCOHOL DEHYDROGENASE
active site

THR

CYS

CYS

LEU LEU

HIS

PHE PRO

PRO

HIS

The hydrophobic Amino acids
keeps the inside of the Enzyme
water-free.

The hydrophilic Amino Acids
are reacting at the surface
of the Enzyme.

Hydrophilic AA with active Groups
at the Peaks of the Pentagramm
1. SER, THR   OH
3. GLU, ASP   COOH
5. ARG, TRP   NH2, NH
7. CYS, LYS   SH, NH3
9. HIS, TYR   NH, OH

Hydrophobic constructive AA at
the Low Points of the Pentagramm
2. GLY, PHE
4. ALA, ILE
6. PRO, VAL
8. GLN, ASN   NH2 Transport
0. MET, LEU

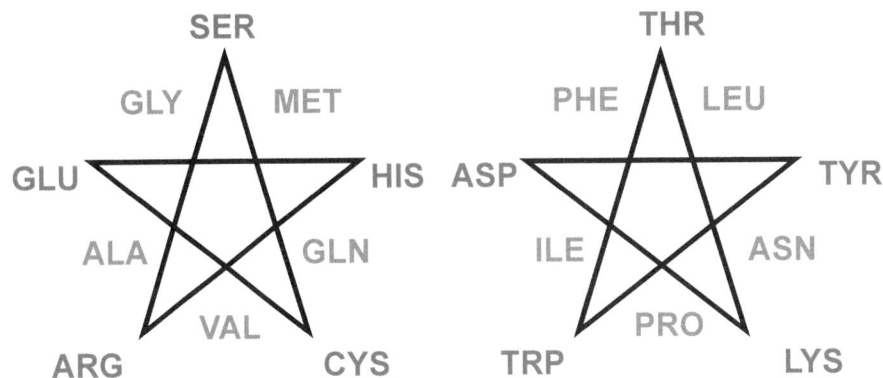

AA - parallels are like in the music minor-parallels (in dales) to the major chords (at the peaks of pentagram).
AA at the peaks produce specific active product:

1 SER   ethanolamine, choline
    4 ALA   acetate
3 GLU   transamination
    6 VAL   isobutyrate
5 ARG ornith.,creatine,vasopres.
    8 GLN   NH2- transport
7 CYS   taurine, oxidoreductions
    10 MET homocysteine
9 HIS histamine, purines
    2 GLY   purinebases

THR   propionate, B12
ILE   propionate
ASP transamination
PRO aminovalerate
TRP B-vit., nicotinacids
ASN   NH2-transport
LYS   vasopres., acetoac.
LEU   acetoacetate
TYR   tyramine, adrenaline
PHE   phenylacetate

In the right pentagram are amino acids predominantly ketoplastic.

Distance of planets from the sun and distance of tone C in music octaves

☿	♀	⊕	♂	Asteroids	♃	♄	♅	♆	♇	
3.9	7.2	10	15.2	29	52	95	192	301	395	- observed
0	3	6	12	24	48	96	192		384	- calculated

Frequencies (Hz) of the octaves of sound C:

C2	C1	C	c	c′	c″	c‴	c⁗	c′′′′′	
16	33	65	131	262	523	1040	2093	4186	
1	2	4	8	16	33	65	130	261	- relative to C2

According to the rule of Titius (1729-1796) and Bode (1747-1826) observed distance planets from the sun relative to the distance of Earth = 10, minus 4, gives the calculated distance of planets. The distance of next planet is double of the preceding planet.

Also the frequency in Hertz between the tones C through eight octaves is double of the preceding tone C.

E-58

Recalculation of music scales to 360 grade circle

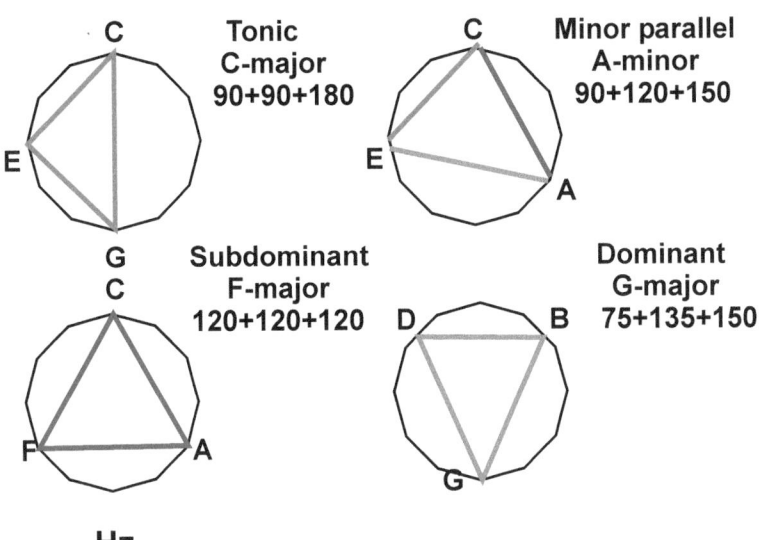

Tonic
C-major
90+90+180

Minor parallel
A-minor
90+120+150

Subdominant
F-major
120+120+120

Dominant
G-major
75+135+150

	Hz	
C`	261	0
Db	277	15
D	294	45
Eb	311	60
E	330	90
F	348	120
Gb	370	150
G	392	180
Ab	415	210
A	440	240
Bb	466	300
H(B)	494	330
C``	522	360

Every scale must be separately
recalculated to 360 grade circle.
Then every tonic, subdominant
and dominant looks identical.
The example presented here is
for C-major scale.

In the zodiac are the temperaments in following succession: fire, earth, air, water.
The temperaments fire and air are masculine active, the temperaments earth and water are feminine restorative.

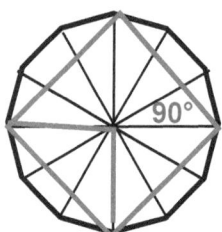

Square 90 grade = 3 x 30 grade
Tension between temperaments: fire - water, water - air, air - earth, earth - fire.
These people are active up to the high age.

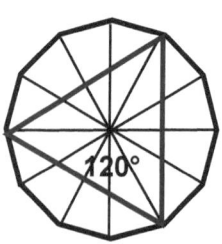

Trine 120 grade = 4 x 30 grade
Monotone, because is connecting the same temperaments: fire - fire, earth - earth, air - air, water - water.
These people gain results without effort, later can become lazy.
Trine in music is the subdominant.

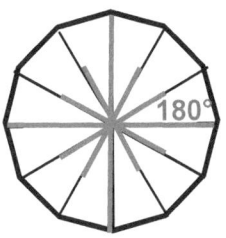

Opposition 180 grade = 6 x 30 grade
Polarity of complementary temperaments
fire - air, earth - water.
This aspect cause essential changes.
One opposition and two squares
correspond to tonic in music.

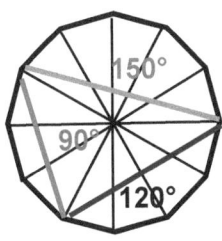

Quincunx 150 grade = 5 x 30 grade
Tension male - female:
fire - earth, air - earth, fire - water,
air - water. This aspect is provocative.
In music combination of trine, quadrate
and quincunx correspond to minor
chord.

Transformation of music chords to the pentagram

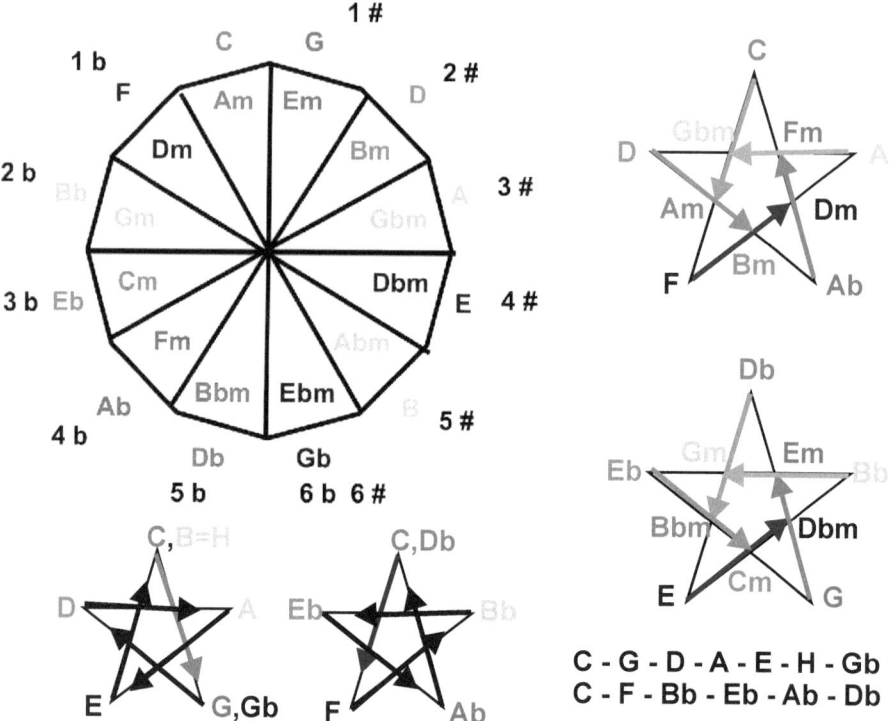

C - G - D - A - E - H - Gb
C - F - Bb - Eb - Ab - Db

Transformation of the major chords from circle to the pentagram, in direction to right C-G-D-A-E-B-Gb, in direction to left C-F-Bb-Ab-Db. The minor chords having the same name as major chords are in opposition in the pentagram. The result are the minor parallels, three steps (thirds) in direction to the left.

E-62

Performance triangles: F-major, C-major

Tonic	F-major	C-major
Minor-parallel	D-minor	A-minor
Subdominant	**Bb-major**	**F-major**
Minor-parallel	G-minor	D-minor
Dominant	C-major	G-major

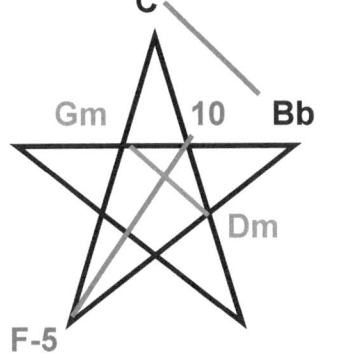
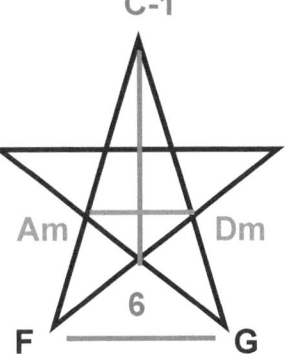

Placement of chords in the pentagram gives the possibility to illustrate harmonic connection of chords in a specific scale.

E-63

Comparison of the frequency of tones with the frequency of colors of chakras I to VII

G´	A´	B´	C´´	D´´	E´´	F´´	
392	440	494	523	587	659	698	Hz
red	orange	yellow	greenn	blue	indigo	violet	
400	450	500	550	600	650	700	x10 ↑ 12Hz
I	II	III	IV	V	VI	VII	

red 400 + violet 700 = 1100
orange 450 + indigo 650 = 1100
yellow 500 + blue 600 = 1100
mean 1100 / 2 = 550 green

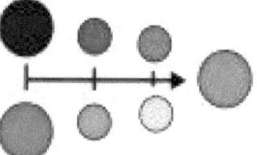

Frequency of tones is compared with frequency of colors. First chakra I at the back basis of body has red color, the tone is G. Chakra VII at the head has violet color and the tone is F. Chakra IV at the middle of body is green and the tone is C. Green is the mean of the shown pairs of color.

On the basis of this knowledge is on the next page the music pentagram placed on the body, with chord C-major with green color on the heart at the front of the body.

Major and minor chords on the body projected using the music pentagram

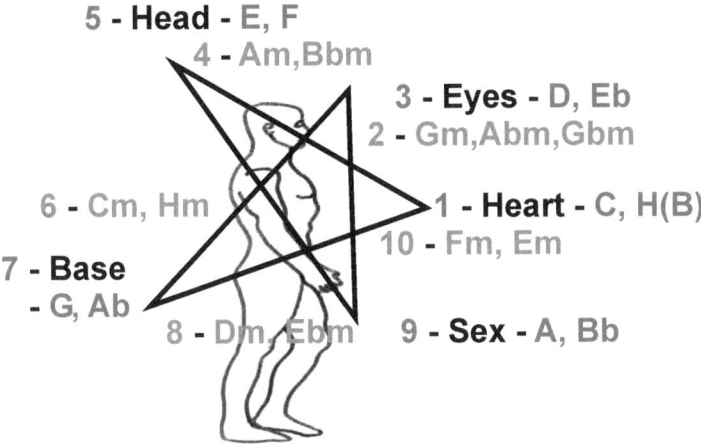

5 - **Head** - E, F
4 - Am,Bbm
3 - **Eyes** - D, Eb
2 - Gm,Abm,Gbm
6 - Cm, Hm
1 - **Heart** - C, H(B)
10 - Fm, Em
7 - **Base**
- G, Ab
8 - Dm, Ebm
9 - **Sex** - A, Bb

At peaks of the pentagram are major chords, in dales of the pentagram and in the inside of the body, are minor chords.

E-65

Song: C-Major

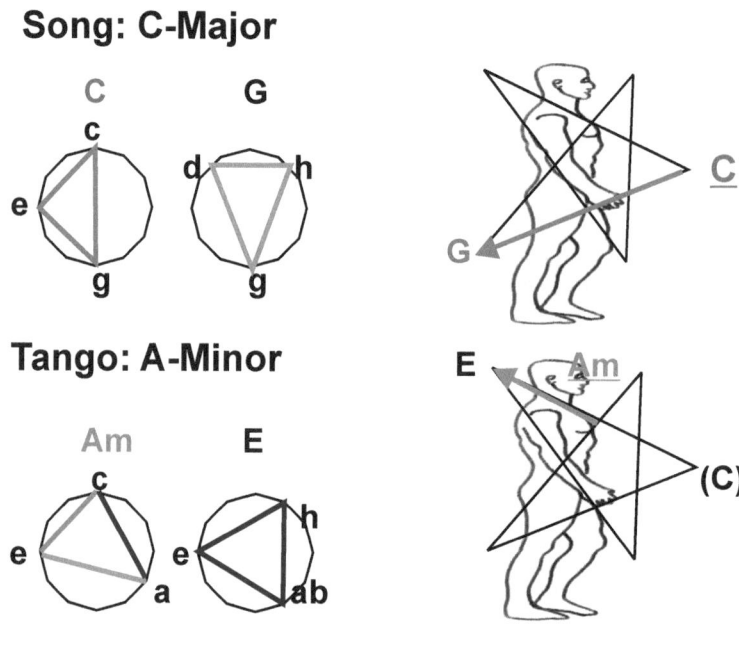

Tango: A-Minor

Parallel: C-Major → A-Minor

In the song C-major the energies move between middle of body at the front to back of the body at bottom, what support the movement of legs.

In tango Ole guapa of Malando the movement of energies of the minor chords is at the head, what supports the head to head hold of the partners by argentine tango.

Method for forecast of weather

1. The Earth move on the ecliptic during the year and around own axis during 24 hours. The cross point of the ecliptic and of the equator in the spring (0 grad Aries) determine division of 360 grad circle to 12 sectors. Heliocentric aspects of a planet with Earth activate its action. Geocentric axis AC-DC (rise and setting of sun, moon and planets) and MC-IC (their culmination at noon and midnight) projected by to world map with Astro - Carto - Graphy shows the places, where the action of planets come to expression.

2. The wind of mercury is at the quarter of moon stormy, if mercury is activated by heliocentric aspect with Earth. Direction of the movement of hurricanes, typhoons and tsunami can be influenced by mercury, if his line is close to the area and his action is activated by aspect with Earth. Saturn with his north streaming brings coll weather, Mars hot and dry weather, Venus moist, Jupiter cumulus clouds, Pluto rainfall, Neptune high water, Uranus high pressure.

3. Positions of planets were calculated at new moon and full moon for GMT (Greenwich), additionally also 6 fays later, but with the same time as by full or new moon. Heliocentric and geocentric positions of planets and Astro - Carto - Graphy can be calculated by software WinStar. The moon positions can be found by software of Peterhans (1998) for the time from -1400 to +2200.

Method of long-term forecast of weather

Heliocentric   Geocentric   Astrokartography

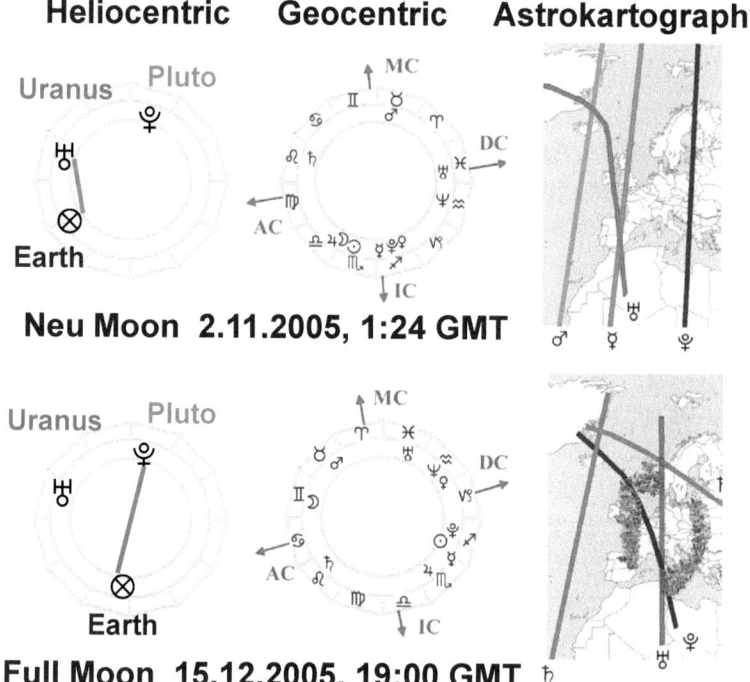

Neu Moon 2.11.2005, 1:24 GMT

Full Moon 15.12.2005, 19:00 GMT

On 2.11.2005 Pluto was heliocentrically not in aspect with earth = no rain.

On 15.12.2005 Pluto was heliocentrically in opposition to earth = rainfall, but Uranus divided it into two parts.

E-68

Opposition action of planets

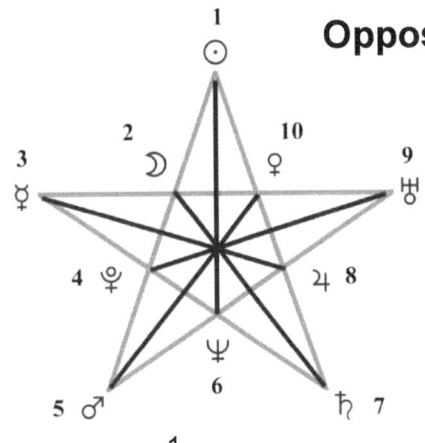

Uneven male numbers: **expansion**
Even female numbers: **connection**

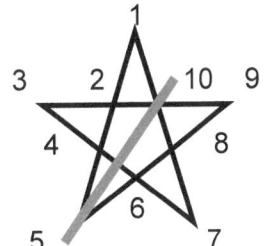

Triangles to axis
5-Mars - 10-Venus

1-Sun	- 9-Uranus :	clear
2-Moon	- 8-Jupiter:	cloudy
3-Mecury	- 7-Saturn:	streaming
4-Pluto	- 6-Neptune:	humid

1-Sun - **clear, warm**
6-Neptune - **fog, high water**
3-Mercury - **wind, storm**
8-Jupiter - **quiet, cumulus clouds**
5-Mars - **dry, hot weather**
0-Venus - **moist, good for vegetation**
7-Saturn - **north streaming, cold**
2-Moon - **change of weather**
9-Uranus - **high pressure**
4-Pluto - **rainfall**

**New Moon 8.8.2002,19:14 GMT
Flood 2002 in Europa**

⊗ Earth ♅ Uranus ♆ Neptun ♇ Pluto

**14 days rainfall on one place started on August 7, 2002
at the crossing of Pluto with Uranus on 9E/49N. The
furrow of high pressure of Uranus hindered movement
of rain clouds of Pluto.**

E-70

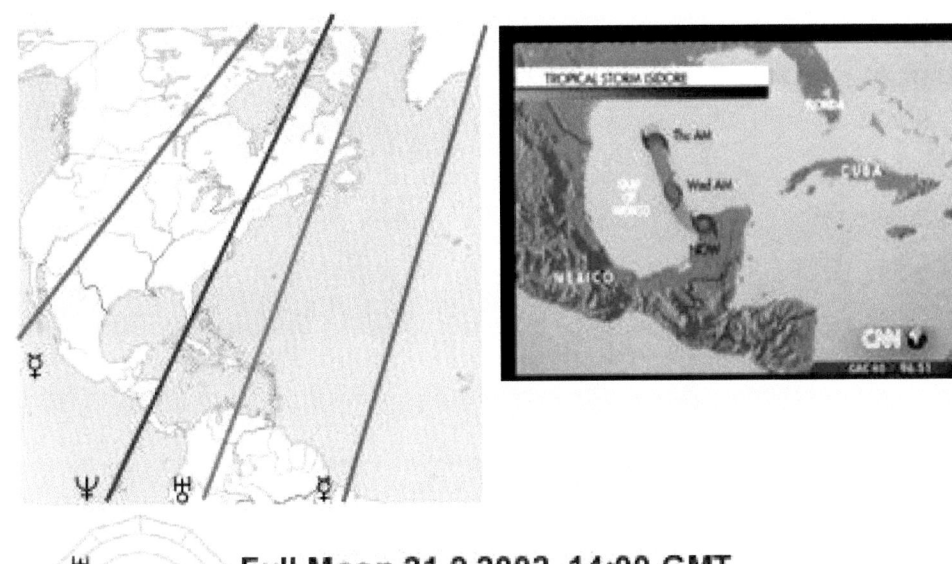

Full Moon 21.9.2002, 14:00 GMT
Hurrikane Isidora

⊗ Earth ☿ Mercury ♅ Uranus ψNeptun

On full moon was Mercury heliocentrically in conjunction with Earth. Mercury from both sides moved the hurricane Isidora to the west. Damages were in Mexico and Louisiana.

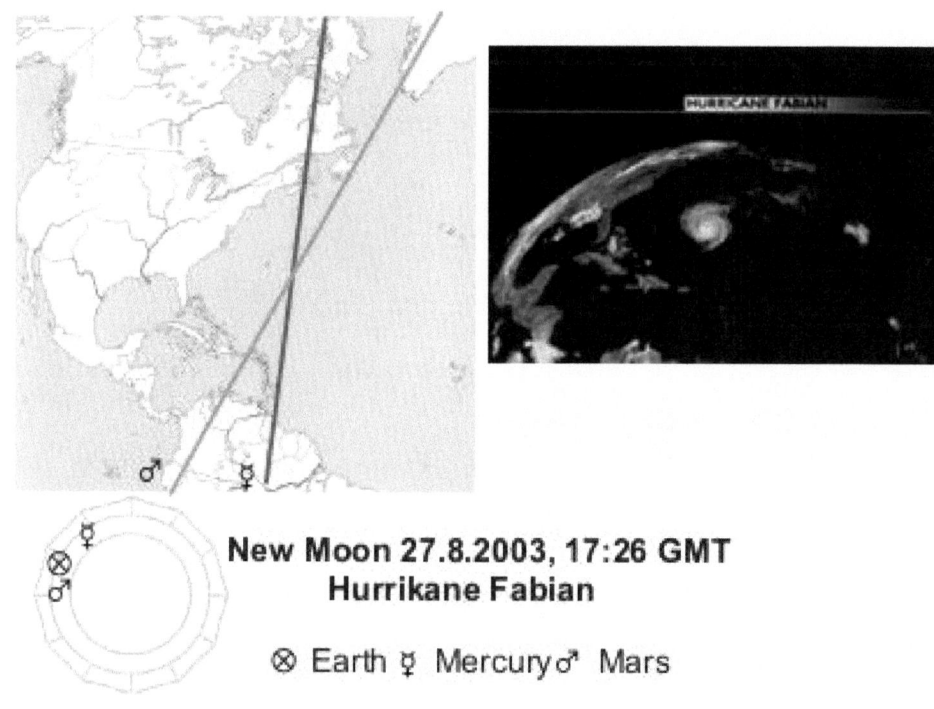

New Moon 27.8.2003, 17:26 GMT
Hurrikane Fabian

⊗ Earth ☿ Mercury ♂ Mars

Heliocentrically was only Mars in conjunction with Earth.
Mercury was not in aspect with Earth. Hurricane Fabian
remained standing at Bermuda.

E-72

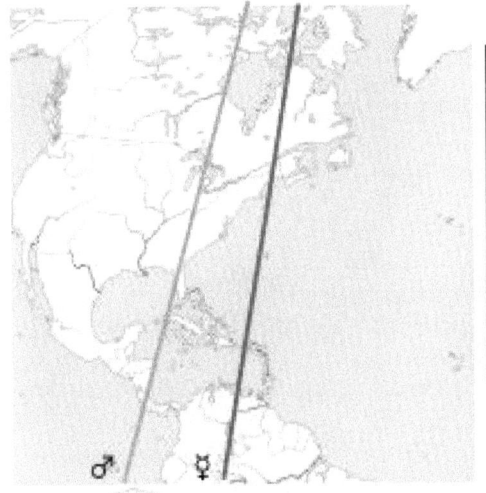
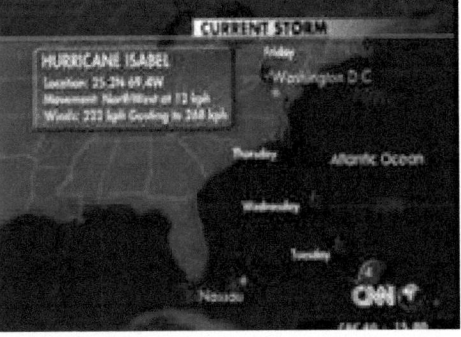

Full Moon 10.9.2003, 16:37 GMT
Hurrikane Isabel

⊗ Earth ☿ Mercury ♂ Mars

On full moon were Mars and Mercury heliocentrically in conjunction with Earth. Hurricane Isabel made damages in Maryland and Virginia.

E-73

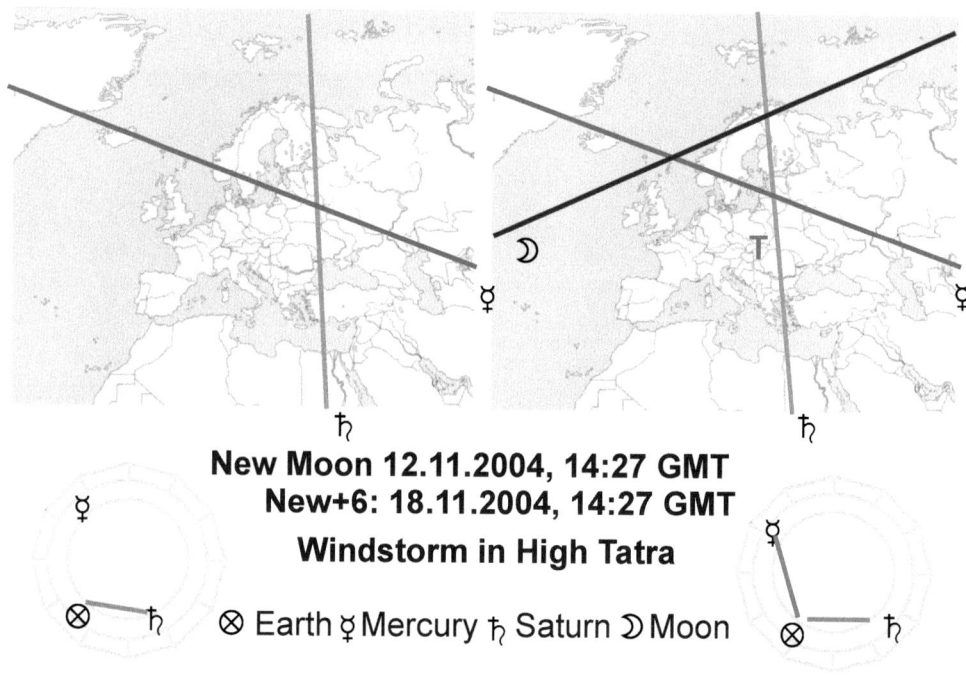

New Moon 12.11.2004, 14:27 GMT
New+6: 18.11.2004, 14:27 GMT
Windstorm in High Tatra

⊗ Earth ☿ Mercury ♄ Saturn ☽ Moon

On new moon was Saturn heliocentrically in aspect with
Earth and with his north streaming made damages in
Poland.
6 days after new moon was Mercury heliocentrically in
exact square with Earth and together with Saturn totally
damaged forest in High Tatry (T) in Slovakia.

E-74

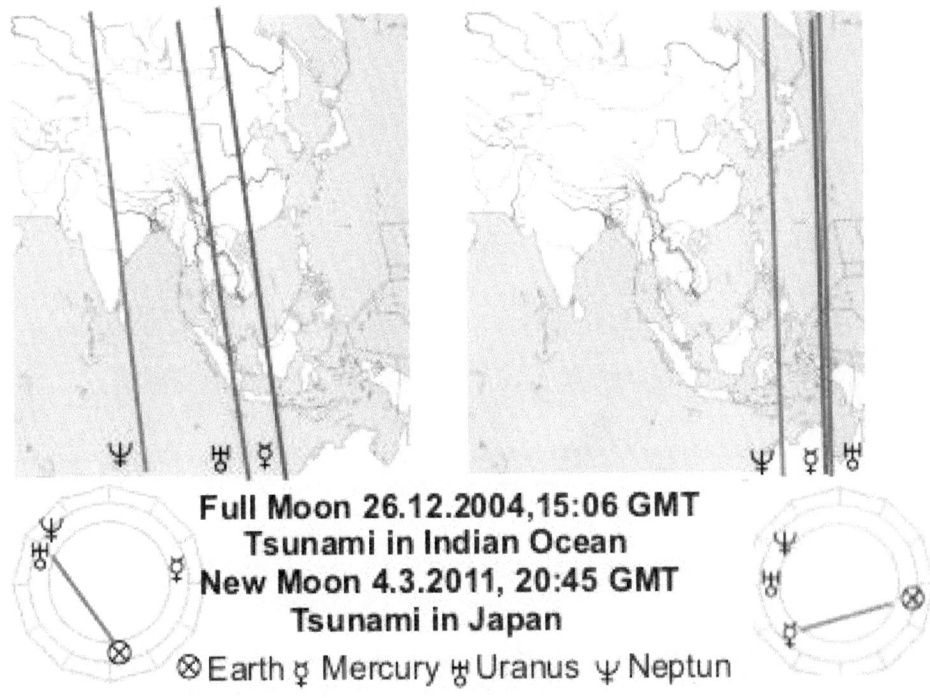

Full Moon 26.12.2004,15:06 GMT
Tsunami in Indian Ocean
New Moon 4.3.2011, 20:45 GMT
Tsunami in Japan

⊗ Earth ☿ Mercury ♅ Uranus ♆ Neptun

At the time of Tsunami in Indian Ocean was Uranus close to the location of the earthquake. Mercury (wind) was eastward, Neptune (water) was westwards of the place of earthquake.
At the time of Tsunami in Japan was the situation similar: Mercury was eastward, Neptune westward.

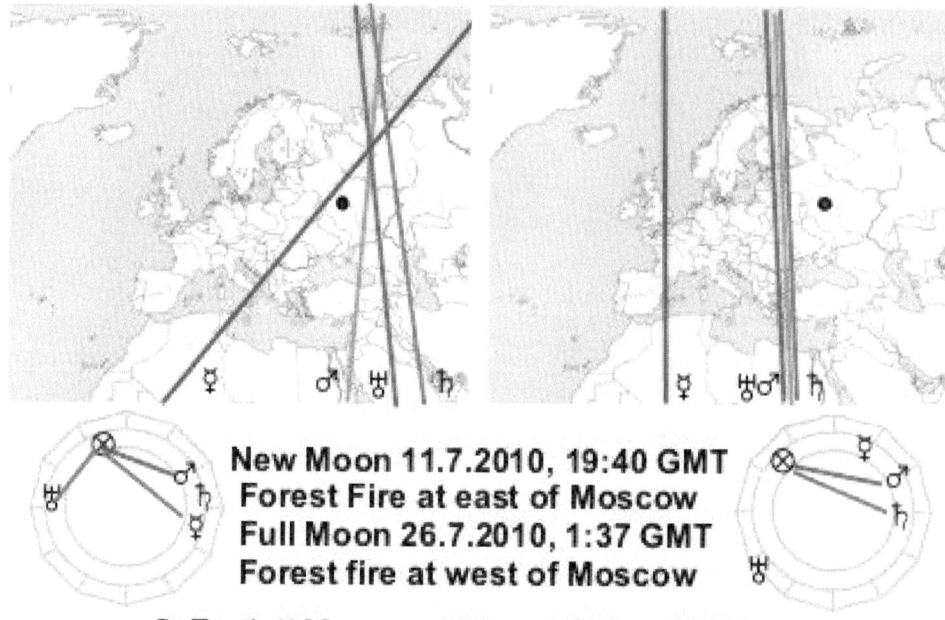

New Moon 11.7.2010, 19:40 GMT
Forest Fire at east of Moscow
Full Moon 26.7.2010, 1:37 GMT
Forest fire at west of Moscow

⊗ Earth ☿ Mercury ♂ Mars ♄ Saturn ♅ Uranus

On new moon were heliocentrically Mars, Uranus and Mercury in aspects with Earth. In entire Europa was hot weather, eastward of Moscow started forest fire.

On full moon was Mars still heliocentrically in square with Earth. Geocentrically was Mars now westwards of Moscow, where new started new forest fire.

E-76

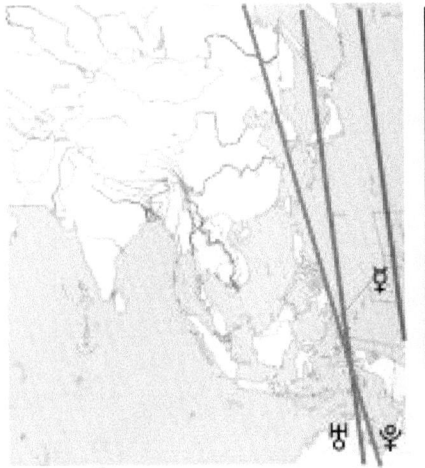

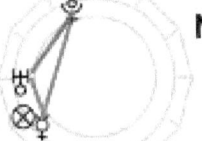

New Moon 3.11.2013,12:49 GMT
Taifun Haiyan

⊗ Earth ☿ Mercury ♅ Uranus ♇ Pluto

By Typhoon Haiyan in Philippines heliocentrically were Mercury, Uranus and Pluto in aspects with Earth. Uranus and Pluto were crossing at the place where Typhoon started. Wind of Mercury moved it to Philippines.

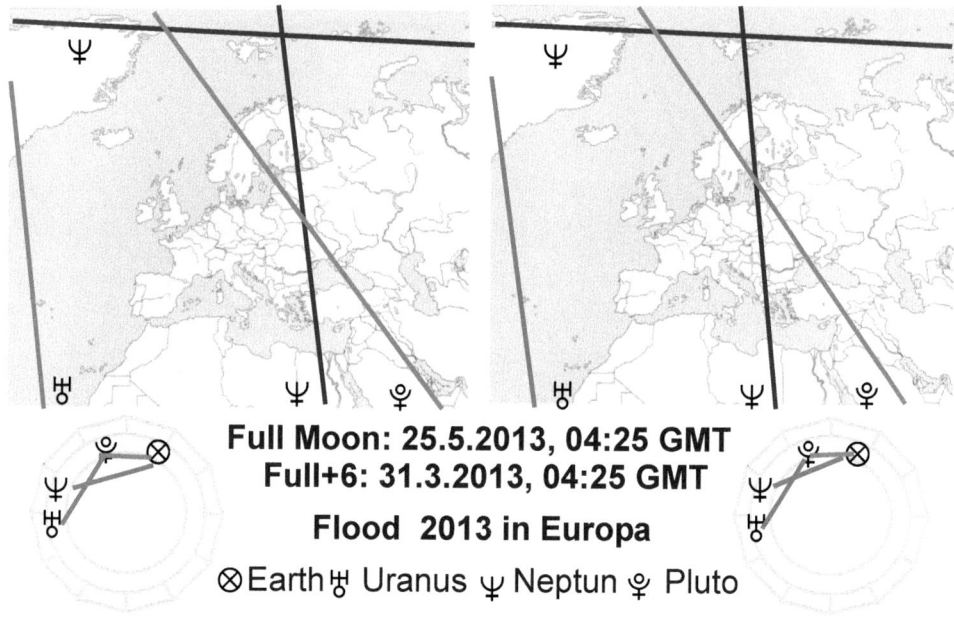

Full Moon: 25.5.2013, 04:25 GMT
Full+6: 31.3.2013, 04:25 GMT

Flood 2013 in Europa

⊗ Earth ⛢ Uranus ♆ Neptun ♇ Pluto

Neptune with Pluto caused at there crossing heavy rain in the east Germany and in Bohemia.
The rain started 6 days after full moon. Neptune, Pluto and Uranus were heliocentrically the whole time in aspects with the Earth.

E-78

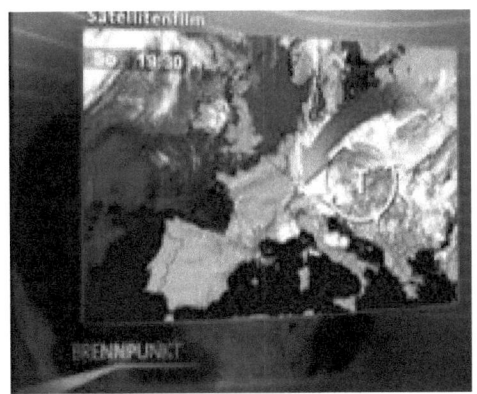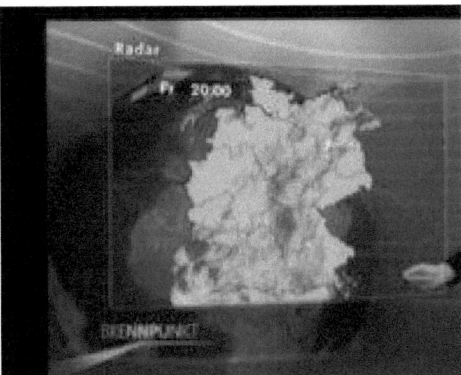

Full moon 25.5.2013:

The huge quantity of water caused the crossing of two
lines of Neptune. The vertical line is MC-Neptune, the
horizontal line is AC-Neptune.

At the crossing with DC-Pluto line started huge rainfall
6 days after full moon over Baltic sea. The satellite
pictures demonstrate it.

E-79

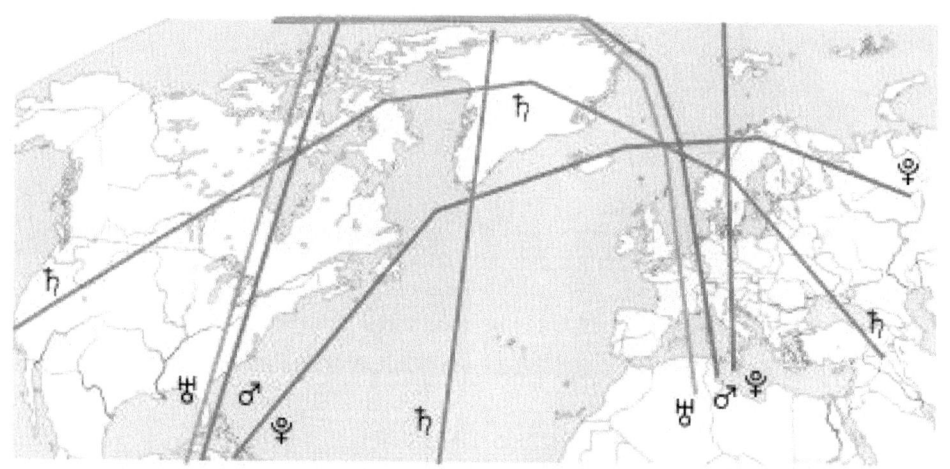

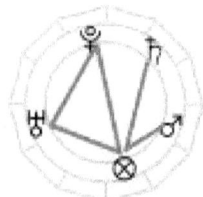

New Moon 1.1.2014, 23:14 GMT

Niagara Falls frozen in 2014

⊗ Earth ♂ Mars ♄ Saturn ♅ Uranus ♀ Pluto

After new moon on January 1, 2014 was in the USA cool, in Europe warm weather. In the USA were effective MC-Uranus and IC-Mars with they north streaming, together with cold streaming of AC-Saturn from the crossing with MC-Saturn over Greenland. In Europa were effective AC-Uranus and DC-Mars without north streaming.
6 days later were the lines more westwards. On January 9 frozen the Niagara Falls.

E-80

Periods of the world history

The periods of the world history are labeled by three numbers: 600-,60-,6- year period. The beginning was set to transition from hunting to agriculture with year -4477. After ten 600- year periods new beginning period falls to year 1523. First will be shown 600- year periods with even constructive female periods, then uneven active male periods. In tables at the end of the book are the period / year equivalents presented.

Even periods: connection, synthesis, construction

000 (-4477): agriculture, calendar
200 (-3227): wedged writings, observation of stars
400 (-2077): calculations of the Summers
600　(-877): philosophy of the Greeks
800　(+323): first ecumenic council
000 (+1523): reformation, human rights

Uneven periods: expansion, activity, fight

100 (-3877): construction of cities in Creta
300 (-2677): central might of pharao in Egypt
500 (-1477): conquests: Assyrians, Egyptians
700　(-277): conquests: Romans
900　(+923): conquests: Arabs, Turks
100 (+2123): ???

60- year Periods with No. 5 = Decomposition, Fight

250	-2977	Egypt with Pharao as a head and commander
350	-2377	Dynasty of princes gain the power in Egypt
450	-1777	The follower crisis in Egypt
550	-1177	Tribe Levi takeover authority in Jerusalem
650	-577	Patrician and plebeian polarize in Rome
655	-547	Sparta founded Peloponian defending union
750	23	Jesus of Nazaret active since year +30
755	53	Empirer Nero (since 54) burned in 64 Rome
850	623	Mohammed captured Mekka
855	653	Arabs gained Syria, Mesopotamia, Lybia
950	1223	Mongolian spread to the west
955	1253	Pape Innocenc IV. permitted the torture
050	1823	Liberation of Mexico, Peru

60- year Periods No. 9=Creation, 0=Start, 1=Identity

590	-937	Colonization of Greek Isles
690	-337	Alexander Great defeated Persian King
790	263	Constantin Great defeated Rivals
890	863	Orthodox (Methodius) to Moravia
990	1463	Humanism in Europa
600	-877	Assyrian subjugated Phoenician
700	-277	Expansion of Rome
800	323	1st ecumenical council
900	923	Renaissance of German King Otto
000	1523	Reformation of Christian Church
610	-817	Instead of wars - 1st Olympic Games
710	-217	Hannibal crossed the Alps with elephants
810	383	2nd ecumenical council, Christ. State Religion
910	983	Magyars baptized, Stephen - Hungarian King
010	1583	Counter - reformation (since 1596)

USA: 6- year Periods 1=Identity, 4=Demarcation

6- year period No. 1 = Identity
041: 1769 Resistance against colonial command
051: 1829 Monroe doctrine about whole America 1823
061: 1889 Spanish-Amerikanish war about Cuba 1895
071: 1949 Truman doctrine about cold war East-West
081: 2009 Obama make politics for own citizens

6- year period No. 4 = Demarcation
044: 1787 With fundamental law started USA exist
054: 1847 War USA-Mexiko 1845-1848 about California
064: 1907 Republic Panama with help of USA founded
074: 1967 Vietnam war, Breshnev doctrine about blocks

USA: 60- year periods 7=Turn, 8=Equilibrium

6- year periods
070: 1943 Division of world to East- and West-Blocks
071: 1949 Korea war 1950 - 1953
074: 1967 Vietnam war 1965 - 1973
075: 1973 Helsinki treaty about cooperation of two blocks
            Shift after 30 years:
076: 1979 Reagan started armament race, then
077: 1985 Reagan with Gorbacov relaxed the politics
078: 1991 1st Irak war, end of Soviet Union
079: 1999 Intervention in Jugoslavia finished the war

080: 2003 War in Afghanistan and 2nd war in Irak
081: 2009 Upheavals in arabic countries (2011)

Russia: Rise and fall of communism

1st 60- year period correspond No. 6 = construction
6- year periods
060: 1883 **0=Start**   Industrialization in Europa
061: 1889 **1=Identity**   Marxism came to Russia
062: 1895 **2=Realization** Social-demokrat. Party (1898)
063: 1901 **3=Conditions** Bolshevik separated (1903)
064: 1907 **4=Demarcation** Czar Nicholas founded duma
065: 1913 **5=Decomposition** World war, 1917 revolution
                Shift after 30 years:
066: 1919 **6=Construction** Soviet Union with Stalin (1922)
067: 1925 **7=Turn** Collectivization, persecutions (1927)
068: 1931 **8=Equilibrium** Leading role of the party
069: 1937 **9=Creation** 2nd world war with modern weapons

From **1567** when Ivan IV terrible persecuted bojars, to
1**927** when Stalin persecuted kulaks, elapsed **360 years**

2nd 60- year period correspond No. 7 = turn
6- year periods
070: 1943 **0=Start**   Turn of war in Stalingrad
071: 1949 **1=Identity** Communism to East Europa
072: 1955 **2=Realization** Destalinization by Chruscow
073: 1961 **3=Conditions** Wall in Berlin constructed
074: 1967 **4=Demarcation** Czechoslovakia (CS) occupied
075: 1973 **5=Decomposition** Democracy in CS suppressed
                Shift after 30 years:
076: 1979 **6=Construction** Pope started his travel
077: 1085 **7=Turn**   Gorbacov started new politics
078: 1991 **8=Equilibrium**   End of Soviet Union
079: 1997 **9=Creation**   Reforms in East Europe
080: 2003 **0=Start**   East European states to Europ.Union

The European Union

074: 1967 **0=Start**   Founding of European Union(EU)
075: 1973 **1=Identity**   Extending with Danmark, Ireland, UK
076: 1979 **2=Realization** 1st direct elections to EU parliament
077: 1985 **3=Conditions**   Opening state borders
078: 1991 **4=Demarcation** Unification of two German states
079: 1997 **5=Decomposition**   Reforms in East Europe
              Shift after 30 years:
080: 2003 **6=Construction** East European States to the EU
081: 2009 **7=Turn**   Economic crisis, looking for new rules
082: 2015 **8=Equilibrium**   Equilibrium between the states ?
083: 2021 **9=Creation**   Defining new strategies
084: 2027 **0=Start**   Start of a new 60- year period
088: 2051 **4=Demarcation**   New border regulations ?

Founding of German Reich: 8 = Equilibrium

048: 1811 - in 1815   Das Deutsche Reich
                 as union of free countries
058: 1871 - Das Kaiserreich (Wilhelm I.)
                 with Bismarck as Minister
068: 1931 - Hitler strived for the Third Reich

078: 1991 - Unification of two German states

088: 2051 - in this year will be EU in **4=Demarcation**
              could it mean "United European States" ???

Reformation of Religions

Christ: 1523 (reformation and start of human rights)
from 23 (persecution) to 1523 = 1500 years = 60 x 25

Jew: 1583 (120 years after end of pogrom and
120 years before end of the ghetto)
from 83 (diaspora) to 1583 = 1500 years = 60 x 25

Islam: 2123 (reformation as an end of conflicts
between the variety of groups)
from 623 (capture of Mekka) to 2123 = 1500 years

Hypothesis

Reformation occur after 60 x 25 years
    by Christ from year 23 (persecution)
    by Jew from 83 (diaspora)
    by Islam from 623 (capture of Mekka)
    25 is the sum of all uneven numbers

$$25 = 1 + 3 + 5 + 7 + 9$$

Content of period 090: 2063 is 9=Creation. It could be a *political solution* "New Regionalism".
New Regionalism after Pott (2005) is determined only by space, not by state governments. That means free movement, settlement and employment.
Content of the new 600- year period is 1=Identity. It could mean an *ideological solution* for global identity in form of ecumenical unity of all religions. The basis for it could be the acknowledgment of the Synonyms for God, explained on the next page.

Synonyms for God as a basis for ecumenical unity of all religions

For variety of religions God is the highest principle of being. Every religion has its own concept of God, which is in the centre of the teaching of the religion. Usually it is one of the synonyms for God.

1. **Principle** *Buddhism and Taoism* knows God not as subject or person, only as a principle

2. **Spirit** *Mysticism* knows God as light or spirit or infinite reality

3. **Soul** *Shamanism* practise getting the soul out of the physical body

4. **Person** *Judaism, Christianity and Islam* adore a personal God, who reward or punish

5. **Life** *Hinduism* knows *Brahma* - Creator, *Vishnu* - preserver, *Shiva* - destroyer

6. **Truth** *Jesuits and Moslems* fight for truth, dogma and prescriptions

7. **Love** Religion of the *Sermon on the Mount* is an individual way to God, because who practise love, fulfills all

Comparison of Behavior: Christian / Islam

1. Trinity:

Koran 4:171 - The Messiah Jesus was only minister of God und of his Word.
Koran 5:73 - Unbelievers are who talks: Allah is the third from three.

When the officials of Roman empire converted to the Christianity, they organized church using the roman standards. With trinity they put borders to other churches. Also the east Christ church does not accept the trinity.

2. Treating the women:

Koran 4:34 - The men stand above the women, because Allah gave advantage to one part. The honest women are therefore submissive to God and keep the concealed for themself, because also Allah keep it for him.
Koran 2:228 - The women have the same rights as the men have over them, despite of it the men have the priority. Allah is omnipotent, wise.

Consequently are the women spiritually equal like the men, but to the outside have the men priority. In Islam ere women after Koran subordinated, the Christian practice it voluntary, when man head of the family is, because he nourish the family.

3. Christian and Islamic legislation

In Christian countries are men who rape women punished, in Islamic countries are raped married women punished. Man who did rape will not be punished. So are the women forced dress cover.

Koran 24:2 Woman and man, who are guilty of adultery or prostitution, punish them by hundred strikes.
Koran 24:2 And these who chaste women accuse, but not four witnesses supply - punish them by eighty strikes and let their assertion never be valid.
Koran 24: 31 Speak to the faithful women, that they should cast down they eyes and guard their chastity and their attraction not to show bear, no more show than is necessary and that their scarf pull down from neckline to slash of their dress and that they show to nobody their appeal with exception of their husband or parents ...

4. Killing of the faithful people

Koran 4:92 No one faithful is permitted to kill other one faithful man, if not happen through oversight.
Koran 4:93 and who kill one faithful intentionally, his reward is the hell, where he should remain.

Relations of Islam to Jews and Christian

Koran 5:82 You will surely find, that between all people
the Jews and these who Gods pictures put to the side,
Are embittered opponents of faithful. And you will find
that these who speak we are Christ, to the faithful most
friendly facing one another are. It is therefore, because
between them God learned and priests are and because
they are not haughty.

World claim of Islam

Koran 48:28 He it is who his envoy sent with the leading
and with the religion of Truth, that he make you victorious
above all other religions.

Themes for ecumenical unity of all religions

1 All people are spiritually in principle perfect
2 Men and women have the same rights
3 Women should be protected against men
4 Terrorism and suicide are detestable
5 Isolate but do not fight the terror groups
6 Legislation in the whole world should be unified
7 All communities should rise and cooperate
8 Equilibrium of groups inside of large units
9 Creative competitions between young people
0 Universal wholeness through 10 synonyms of God

This numbering follows the method of steps: 1.Identity,
2.Realization, 3.Conditions, 4.Demarcation, 5.Decomposition,
6.Construction, 7.Environment, 8.Equilibrium, 9.Creation,
0.Wholeness.

Preview of historical development

Preview does not mean prediction what will happen, but what will be the content of the next periods.

There were 600-, 60- and 6- year periods defined. As start was taken transit from hunting to agriculture in the year -4477. After ten 600- year periods fell the first new 600- year period on the year +1523 (0=Start). It is period of reformation, of human rights and of globalization. In the next 600- period (1=Identity) from 2123 will variety of human groups fight for their identity and their identity will be new defined.

Periods which correspond to even numbers are female, synthetic. Periods with uneven numbers are male, active and could be also realized with fight. After present female period 080 will follow male period 090, in which should be found global solutions for new regionalism and human rights..

At present period 080=Equilibrium runs at many places of the world independency fights. This urge us to think, what models should be developed in the period 090=Creation. One approach is to let participate all groups in one government. It is not so simple. Second approach is to give for variety of groups freedom but their relations to the wholeness must be exactly defined. New models must be developed and their realization first in thoughts proved.

Existing states will not change their borders without fight. The independent parts must be integrated into the whole using new models. As an example could the European Union be used. The more developed states solve also integration of migrating people. Similar models should be developed also for other parts of world. Local authorities should not only keep the borders and safety, but also develop conditions for keeping the people in the state.

Development in the present 60- year period 8=Equilibrium could looks as follows:

080: 2003 **0=Start**   War in Afghanistan and Irak
081: 2009 **1=Identity** Upheavals in arabic countries

Preview:
082: 2015 **2=Realization**   Polarization, Occupation
083: 2021 **3=Conditions**   Clarification, Negotiation
084: 2027 **4=Demarcation**   Border regulations
085: 2033 **5=Decomposition**   Fight, or Regeneration

Shift after 30 years to new period **9= Creation**

086: 2039 **6=Construction**   Building new Structures
087: 2045 **7=Turn**   New inner Politics
088: 2051 **8=Equilibrium**   New Balance of Power
089: 2057 **9=Creation**   New Strategies
090: 2063 **0=Start** of new **60-** year period with No.
          9=**Creation:** Models for new Regionalism ?

Summary

The ten whole numbers have a spiritual content, which describe spiritual and natural processes. The definition include also ten synonyms for God. From the synonyms could be everything derived. The synonyms are also basis for keeping health by thoughts. Whole numbers building geometrical figure describe performance triangles. This was demonstrated in psychic, chemical and music examples.

Kepler (1571-1630) showed that also relations between planets are activated when they build heliocentrically an aspect with Earth corresponding to geometrical figure. In this book was shown, that it can be used for long-term predictions of the weather. Geocentric axes AC-DC and MC-IC projected on the world map show where in the world the weather actions of planets are realized.

Personal development occur in 6- year periods. The world history can be described by 600-, 60- and 6- year periods.

600-, 60-, 6- year Periods

600	60	0	1	2	3	4	5	6	7	8	9
6	9	-337	-331	-325	-319	-313	-307	-301	-295	-289	-283
7	0	-277	-271	-265	-259	-253	-247	-241	-235	-229	-223
	1	-217	-211	-205	-199	-193	-187	-181	-175	-169	-163
	2	-157	-151	-145	-139	-133	-127	-121	-115	-109	-103
	3	-97	-91	-85	-79	-73	-67	-61	-55	-49	-43
	4	-37	-31	-24	-19	-13	-7	-1	+5	+11	+17
	5	+23	+29	+35	+41	+47	+53	+59	+65	+71	+77
	6	+83	89	95	101	107	113	119	125	131	137
	7	143	149	155	161	167	173	179	185	191	197
	8	203	209	215	221	227	233	239	245	251	257
	9	263	269	275	281	287	293	299	305	311	317
8	0	323	329	335	341	347	353	359	365	371	377
	1	383	389	395	401	407	413	419	425	431	437
	2	443	449	455	461	467	473	479	485	491	497
	3	503	509	515	521	527	533	539	545	551	557
	4	563	569	575	581	587	593	599	605	611	617
	5	623	629	635	641	647	653	659	665	671	677
	6	683	689	695	701	707	713	719	725	731	737
	7	743	749	755	761	767	773	779	785	791	797
	8	803	809	815	821	827	833	839	845	851	857
	9	863	869	875	881	887	893	899	905	911	917

Example: The year 917 corresponds to the Period 899

E-94

600-. 60-. 6- year Periods

600	60	0	1	2	3	4	5	6	7	8	9
9	0	923	929	935	941	947	953	959	965	971	977
	1	983	989	995	1001	1007	1013	1019	1025	1031	1037
	2	1043	1049	1055	1061	1067	1073	1079	1085	1091	1097
	3	1103	1109	1115	1121	1127	1133	1139	1145	1151	1157
	4	1163	1169	1175	1181	1187	1193	1199	1205	1211	1217
	5	1223	1229	1235	1241	1247	1253	1259	1265	1271	1277
	6	1283	1289	1295	1301	1307	1313	1319	1325	1331	1337
	7	1343	1349	1355	1361	1367	1373	1379	1385	1391	1397
	8	1403	1409	1415	1421	1427	1433	1439	1445	1451	1457
	9	1463	1469	1475	1481	1487	1493	1499	1505	1511	1517
0	0	1523	1529	1535	1541	1547	1553	1559	1565	1571	1577
	1	1583	1589	1595	1601	1607	1613	1619	1625	1631	1637
	2	1643	1649	1655	1661	1667	1673	1679	1685	1691	1697
	3	1703	1709	1715	1721	1727	1733	1739	1745	1751	1757
	4	1763	1769	1775	1781	1787	1793	1799	1805	1811	1817
	5	1823	1829	1835	1841	1847	1853	1859	1865	1871	1877
	6	1883	1889	1895	1901	1907	1913	1919	1925	1931	1937
	7	1943	1949	1955	1961	1967	1973	1979	1985	1991	1997
	8	2003	2009	2015	2021	2027	2033	2039	2045	2051	2057
	9	2063	2069	2075	2081	2087	2093	2099	2105	2111	2117
1	0	2123	2129	2135	2141	2147	2153	2159	2165	2171	2177

Example: The year 2177 corresponds to the Period 109

E-95

Literature used

Assagioli, R. Psychosynthesis, Hobs, Dorman,
    N.Y. 1965
Chia,M., Li, J. The inner structure of Tai Chi,
    Healing Tao Books, Huntington, N.Y. 1996
Coulson, R.A. Metabolic rate and the flow theory,
    Comp.Biochem.Physiol. 84A, 217-229, 1986
Der Neue Brockhaus, F.A.Brockhaus,
    Wiesbaden, 1978-1980
Goodavage, J.F. Write your own horoscope,
    New American Library, N.Y. 1968
Huber B. & L. Lebensuhr im Horoskop,
    Verlag API, Adliswil / Zürich, 1980
Jung, K.M. Weltgeschichte in einem Griff,
    Safari Verlag,   Berlin, 1979
Koch, W.A. Aspektlehre nach Johannes Kepler,
    Kosmobiologische Gesellschaft, Hamburg, 1952
Lewis, J. Astro-Carto-Graphy,
    F.L.Francisco, San Francisco, CA, 1976
Peterhans, E. Aspector-Plus-Software, Chom, 1998
Pott, H.G. Kurze Geschichte der europäischer Kultur,
    W.Fink Verlag, Paderborn, 2005
Reichl, J. Scientific and esoteric Way to the Health,
    Published by the Author, 2006
Reichl, J. The universe is composed of whole numbers,
    Published by the Author, 2011
Reichl, J. You get everywhere with love, Autobiography,
    Published by the Author, 2011
Yücelen, Y. Was sagt der Koran dazu ?
    Deutscher Taschenbuchverlag, München, 1986
WinStar - Astrology Software, Big Rapids, MI, 1998,2009